Florida's Citrus Canker Epidemic: Pieces of a Puzzle

David Lord

Florida's Citrus Canker Epidemic: Pieces of a Puzzle

Photo Credit:

The image of citrus canker as shown on the front cover of this book was obtained from the PaDIL image library at www.padil.gov.au. This image library is maintained by the Plant Biosecurity Cooperative Research Center (PBCRC) in Australia. The image is freely distributed under the Creative Commons Attribute 3.0 Australia. The image is being used solely as part of the description of citrus canker symptoms, and for no other purposes.

ISBN-13: 978-1539131557
ISBN-10 1539131556

Library of Congress Cataloging-in-Publication Data is on file with the Library of Congress.

Printed in the United States of America

Acknowledgements:

I am very grateful for the assistance given to me in the preparation of this book. Jack and Patty Haire were particularly helpful in reading through the initial draft, and making valuable comments. I am very appreciative for the help and encouragement given by Mr. Andrew Meyers, who took time out to read and comment on the initial draft.

All errors and omissions are solely the fault of the author. An errata page will be maintained at www.citruscankerdocs.com and comments are always welcomed.

CONTENTS

Part I: Just the Basic Facts

Part II: Going Beyond the Basic Facts

Author's Note

This is the book that I intended to write for 10 years. It is finally complete. I did not want to summarize the conclusions of others. I had new ideas stemming from my own original research. You can call it an investigation, a quest or an obsession.

This book is about the citrus canker epidemic and the last unsuccessful program intended to effectively eliminate the plant disease from Florida. Citrus canker causes lesions to form on the stems, leaves and fruits of citrus trees. Fresh fruit with lesions is unmarketable, but the citrus can still be used for juice.

Citrus canker is a frustrating disease to eliminate. No flares go off when a tree becomes infected with citrus canker. Symptoms can appear slowly in the new flushes high in the canopy of the tree making the presence of canker elusive. The release of bacteria and later entry into the plant tissues of citrus leaves and fruit require wet conditions coming from rain or irrigation systems.

I was skeptical that South Florida's heavy summer downpours were capable of carrying bacteria far beyond the infected citrus tree as the Department of Agriculture had claimed. However, citrus canker had been found long distances beyond the original discovery. I wanted to know how this could have happen. Thus, I needed to know more about the disease, the infection process and mechanisms of dissemination.

Citrus trees with canker disease may produce less fruit. However, this is likely more of a concern to grove owners who strive to maximize production than to the average homeowners. A mature orange tree may produce in excess of 200 oranges, which can be overwhelming to homeowners, particularly those with more than one tree. The disease is not deadly since no citrus tree has ever died of canker in Florida. It is likely that growers were more concerned with economic harm that could come from trade embargos and quarantines than the disease.

The eradication program was massive. It involved cutting all healthy citrus trees within a specified radius from an infected tree. In the first five years, this radius was 125-ft, and then in year 2000, the radius was changed to 1900-ft. A total of 16.5 million citrus trees were destroyed and more than 1.3 billion dollars were spent. Approximately 87,000 acres (136 square miles) of commercial citrus groves were destroyed. By the end of the program in 2006, citrus canker was still alive and well in Florida. No one could possibly call this a successful program.

The book is divided into Part 1, "Getting the Basic Facts" with Chapters 1 to 6, and Part 2, "Going beyond the Basic Facts" with Chapters 7 to 10. Chapter 1 discusses the eradication program from year 2000 to 2006 and the organizations involved in this effort. The pieces of the puzzle relate to both the epidemic and the underlying research leading up to the 1900-ft policy.

In Chapter 2, the prior attempts to eradicate the disease are examined. The next four chapters delve into other aspects of citrus canker— biology, opposition to the program, legal challenges and select topics in epidemiology. By the end of these six chapters, you can have passed Citrus

Canker 101, with a basic understanding of the underlying facts on the epidemic and the eradication program. Congratulations!

Now, it is time for Citrus Canker 102. The Florida Department of Agriculture and Consumer Services (FDACS or the "Department") justified the program based on a field study which began in 1998. The Department repeatedly told residents that the program came from solid science. I had my doubts, which grew as it became more difficult to get information on this field study.

The results of my investigation into the citrus canker field study is provided in Appendices A to G, as provided in the online website. For the intrepid reader, I suggest the online appendices be read first, before returning to the printed book for Chapters 7 to 10. Chapter 7 summarizes the conclusions.

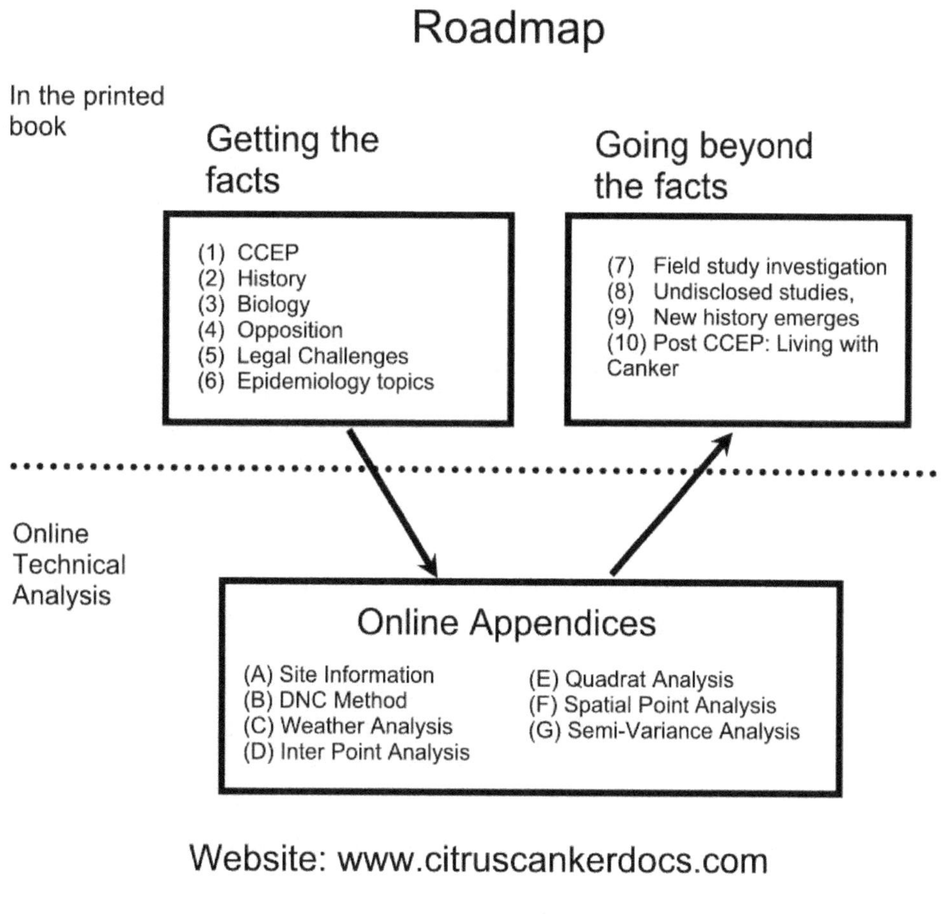

In Chapter 8, a critical piece of the puzzle falls into place. I suggest the true origin of the 1900-ft was not the field study, but a previously undisclosed simulation study. However, the story really comes together in Chapter 9, which provides a new history of the 1900-ft policy and the canker eradication program. It is a sharp contrast to the official narrative created by the Department.

Chapter 10 takes a brief look at the results of living with canker in the post eradication era. The Citrus Health Response Plan (CHRP) seems to be a realistic long term effort to minimize citrus canker dissemination from nurseries to commercial groves. New introductions of canker in groves can be managed through early detection, more frequent fungicide applications and windbreaks. New challenges in citrus disease management lie ahead, particularly with the citrus greening disease, which can kill citrus trees. The efforts to control the future spread of diseases and pests through the Citrus Health Response Plan are commendable.

Admittedly, coverage of some important topics is lacking or only briefly discussed. Economics and politics deserve much more review. The economic impact would include the downfall of the lime industry in South Florida in the 2000 to 2006 period. Millions of dollars were poured into the Homestead area of Miami-Dade County in year 2000, so the citrus industry would have the funds to replant citrus when the disease was gone. However, lime grove owners sold out during the housing boom.

Opponents of the program may be disappointed at how little attention I have given to the self interest of many citrus industry groups, and the allegations of collusion of these interests with government officials. It is natural for plant pathologists from various governmental agencies to work closely with the citrus industry to address their most pressing disease and pest problems. However, this eradication program placed a real burden on residents, who had no voice in the decision making. Most were not aware of the program until it was a "done deal." Their legal recourse seemed as elusive as the citrus canker disease.

Advanced areas of plant pathology are not included in this book. These topics would include microbiology, testing methods, and strain identification. These technical subjects are rapidly evolving, due to collaborative research efforts of scientists worldwide. The sharing of information by US scientists with others, via the Internet, is one reason for optimism in new technological advances. There is hope that new disease resistant varieties of citrus will be developed with DNA research.

What is not lacking, and in fact, is provided in abundance, is supporting documentation in terms of short notes, published articles, minutes of meetings, reports from FDACS and other agencies and news reports, as provided in www.citruscankerdocs.com. These documents provide solid support of the topics in my book and add detailed information, which would not fit within my book. Readers may provide comments and corrections to the topics within my book.

My investigation is done. All topics can not be concluded by a single book. Readers may add their comments to the website and help fill in important areas. Or perhaps, write another book.

David Lord

Part I:

Getting the Basic Facts

1. CITRUS CANKER ERADICATION PROGRAM

> Dear Friends, The Department of Agriculture is currently waging a battle against one of the most devastating, highly contagious diseases known to citrus trees and we need your help... [in the future, after eradication] residents can replant citrus trees that will bear healthy, fresh fruit for themselves, their children and grandchildren to enjoy.

Bob Crawford, Florida Commissioner of Agriculture in promotional brochure on the citrus canker program, distributed in 2000 prior to adoption of the 1900-ft rule.

> Total loss *[from canker]* in citrus productivity is low. The cost of sprays is not high compared to other pests... The most important economic loss is caused by the quarantine restrictions on fruit from citrus canker-infested areas as imposed by citrus canker-free growing countries.

Dr. B.I. Canteros, International Citrus Canker Research Workshop, June 20, 2000. Abstract to presented paper at the conference.[2]

---

1. It's Finally Over, January 10, 2006

A nearly 10 year battle between State of Florida, Department of Agriculture and Consumer Services (FDACS) and a plant bacterial disease known as Asian Citrus Canker had finally ended. The Citrus Canker Eradication Program (CCEP) had resulted in approximately 16.5 million trees destroyed[8] and over 1.3 billion dollars spent.[23,25] However, no declaration of eradication was issued and no celebration marches were in order. The canker bacteria had won. From year 2006 forward, a new approach to the citrus canker disease was adopted, based on control and preventative measures, improved sanitation within nurseries and nursery inspections. Many were elated that the cutting of both healthy and infected citrus trees in residents' backyards was over.

Citrus canker results in unsightly blemishes on the fruit, leaves and stems of citrus trees. Fruit with these unsightly lesions is unmarketable. The citrus fruit quality is generally unaffected by cankers, so the fruit can still be processed for juice. The disease causes only local infections, as the bacteria do not enter the vascular system of the plant. Wet conditions are required for both the release of bacteria and later for the entry of bacteria into new plant tissues.[3] The disease is naturally transmitted to other citrus plants by windblown rain. Images of infected fruit are provided in the online supporting documents website.

The program begun in October 1995 never accomplished its goal of effectively eradicating the citrus canker disease in Florida. On January 10, 2006, USDA Deputy Commissioner Charles Conner gave formal notice that they would longer support funding of the eradication program, because eradication of canker had become infeasible.[30] The Florida Department of Agriculture and Consumer Services (FDACS or "Department") concurred with the decision the following day.[7]

Until today, Department officials have not acknowledged that healthy trees were destroyed. Instead, they have insisted these trees were exposed to the bacteria and would eventually show symptoms of the disease. This position may have been taken for legal reasons, first in response to legal challenges from Broward and Miami-Dade counties and 8 municipalities, and later in response to a class-action lawsuit filed against the Department.

There is no way to know if a citrus tree is infected until lesions appear on the leaves, branches or fruits. Therefore, the trees destroyed by the Department without visible symptoms were indistinguishable from healthy trees. In this book, any tree lacking outward symptoms of canker disease is deemed to be "healthy." An exception to this would be citrus trees that have been inoculated with high doses of bacteria in experimental studies. Also, a tree may be considered exposed or contaminated with citrus canker in retrospect, during the period prior to the onset of disease symptoms.

For years, many Florida residents doubted the merits of the program, feeling the Department had overreacted to the concerns of the citrus industry, and that the program had more to do with trade issues than eradication itself. In the past, Florida banned fruit and other agricultural products from entering the US due to concerns they may carry diseases, so it was anticipated other countries would act similarly if canker was present in Florida.

Opposition was particularly strong in year 2000, when the 1900-ft rule was implemented, as given below. This rule greatly increased the destruction of healthy citrus trees. The rule is, as follows:

> **1900-ft rule:** Every citrus tree within 1900-ft of an infected citrus tree had to be cut down. The rule applied to both residential properties and groves.

Many felt the tree cutting could become more destructive than the disease. Mr. Andrew LaVigne, the Director of Florida Citrus Mutual (FCM), Florida's largest citrus industry organization, stated at the International Citrus Canker Research Workshop in June 2000:

> Many people have mentioned to me, well, where do you stop, you know, you get to the point where the citrus canker eradication program becomes a citrus eradication program. Well exactly. We've got to decide that when we get to that point. Nobody knows what that point is. It is somewhere down the road.[22]

Dr. Timothy Schubert, Administrator of the Plant Pathology Department with the Florida Department of Agriculture and Consumer Services, wrote in 2001, "It was one of the most expensive programs conducted in the world against a plant disease."[26] Yet the program would go on for another five years, and another 800 million dollars would be spent before the USDA decided to quit funding their part of the eradication program.

Dr. Schubert's comments might have sounded a bit familiar to those residents who were around during the last eradication program. At the end of Citrus Canker War II, the Miami Herald, on January 21, 1994, reported, "The ensuing eradication battle — waged in a crisis atmosphere — was one of the most controversial and expensive campaigns over conducted against a citrus pest." The headline on this day was, "All Florida Citrus Trees Declared Canker-Free." Little did they know at the time, that a program at least five times more expensive would be started just 21 months later.

Over time, trees with canker lesions may produce less fruit than healthy trees. Dr. B. I. Canteros, with over 20 years of experience with canker in Argentina, reported the loss of productivity is low.[2,3] Argentina has a long history of living with canker, adopting control measures to minimize its impact. While public relations officers with the Department claimed the citrus trees would ultimately die of canker, they could not produce evidence to support citrus tree death within Florida.

Other countries have cut down both infected and healthy citrus trees including Brazil, Argentina, the Bahamas and Australia. There were however, three very distinguishing features of the Florida program not present in any other country:

- **Level of destruction of uninfected citrus:** The 1900-ft rule resulted in a 260 acre eradication circle. Within this circle, approximately 95% of all citrus trees would be healthy ones, according to the USDA and FDACS reports.[8,29]

- **Clear cutting of residential citrus in several urban areas, including Miami-Dade and Broward Counties:** Making residential areas canker-free seemed synonymous with making them citrus-free. The large areas of citrus-free backyards have been referred to as "clear-cut" areas.[14] The massive destruction of residential healthy citrus trees has no comparison in any of the 26 countries where citrus canker is known to exist.

- **Program cost:** The program cost (1.3 billion dollars or more) is likely far more costly than efforts in other countries. About half of the cost was for growers compensation.[23]

2. We won!

All restrictions on citrus sales were lifted on March 1, 2006. Many residents were overjoyed when local outlets were again allowed to sell citrus trees. The "Citrus for Sale" signs springing up across Florida were victory symbols for those who had worked hard opposing the program. A personal recollection of this sense of joy is provided in Short Note 1.8 on the website.

Miami-Dade and Broward counties and 8 municipalities had challenged the legality of the healthy tree cutting under the 1900-ft rule, beginning in year 2000. The Commissioners from the counties and municipalities of southeastern Florida also approved the actions of their legal departments to challenge the program in court for nearly four years — obviously, no small undertaking. State senators from southeastern Florida opposed the passage of the "new canker law" in year 2002, but it was passed into law. This gave the Department unprecedented authority to search residents' yards using blanket county-wide "agriculture warrants" with probable cause based on a single discovery. The court ruled this law to be unconstitutional, as they violated an individuals' right against unreasonable searches. The program continued for approximately two years after this ruling, by obtaining waivers from residents.

Since March 1, 2006, all varieties of citrus trees can be bought and grown in Florida, which includes Key lime, Persian lime, orange, tangerine, tangelo, grapefruit and lemon trees. Smaller citrus plants such as Key limes, kumquats and calamondins can be grown as bonsai or container plants, if space is limited.

3. The Citrus Canker Eradication – A Brief History

The Citrus Canker Eradication Program (CCEP) was Florida's third attempt to eradicate citrus canker. Canker War I began in 1912 and lasted until 1933. As the canker war was consuming more of the State's budget, it created a minor rebellion in the legislature.[27] Florida was declared canker-free in 1933. Other states were still under quarantine until 1947, when the United States declared the entire country to be canker-free.

Canker War II began in 1984, when scientists thought citrus canker had returned to Florida. It was not citrus canker, but citrus bacterial spot (CBS), a similar looking disease.[26] However, after two years waging war against CBS, true Asian citrus canker was discovered. In 1994, after two years of no new discoveries, the Commissioner of Agriculture declared Florida to be once again "canker-free."

However, this canker-free period lasted only 21 months. A Medfly inspector found canker in a residential backyard in the Sweetwater community, about 5 miles west of Miami International Airport in September 1995. Canker War III can be split into two phases. The first phase (Oct 1995 to Dec 1999) was limited to citrus removals within a radius of 125-ft from an infected tree, and the second phase (Jan 2000 to Jan 2006) greatly expanded this radius to 1900-ft. In the first phase of the program, the Department employed various protocols, including cutting all limbs of the infected trees (hatracking), cutting down only infected trees, and cutting all citrus within 125-ft radius of an infected tree.[26]

Each attempt at eradication proved to be a long and frustrating process. It is likely that Canker Wars I and II went on for far longer and cost far more than had been expected by officials. The focus in this chapter is on Canker War III, second phase employing the 1900-ft policy. The prior attempts, including the first phase of Canker War III are described further in Chapter 2.

Canker War III with the 1900-ft Policy

The 1900-ft rule was implemented by Executive Order by Commissioner Crawford on January 1, 2000. The Department considered it had the legal authority, as provided in the Department's rules and Florida Statutes, to implement the 1900-ft rule in this manner. Official press release of the policy was likely delayed as litigation was ongoing in the Sapp Farms case. On February 11, 2000, Commissioner Crawford announced the 1900-ft policy to the public. In early 2000, hundreds of acres of lime groves in Homestead area of Miami-Dade County were destroyed. By the end of the program, all lime groves in Miami-Dade County, encompassing more than four square miles were gone. Opponents of the program had stated many times that the eradication program was far more destructive than the program, and the disappearance of the lime groves may serve as a good example.

Due to the unprecedented housing boom in South Florida, much of the agricultural land in Homestead and Redland areas of Miami-Dade County has been converted to residential developments. Relaxed lending standards allowed many residents to qualify for subprime mortgages, providing more incentives to convert agricultural land to residential developments. These developments were on the border of the Everglades wetlands, a very environmentally sensitive ecosystem.

Under the 1900-ft policy in residential yards, for every infected tree cut down, on the average approximately 16 to 30 healthy citrus trees would be removed or approximately 95% of the trees were healthy. These estimates are based on a 2002 USDA/APHIS audit by the Office of Inspector General[29] and the 2012 FDACS Comprehensive Report.[8]

The entire program cost has never been formally evaluated. Since a class action lawsuit is still active, additional costs are likely to continue, even 10 years after the last citrus tree was cut. Using cost estimates from an University of Florida/IFAS study[35] and expenditures as reported in the Lakeland Ledger on October 6, 2006, expenditures of 1.2 billion dollars were accounted for the CCEP considering reported USDA and FDACS expenses. The total program cost is at least 100 million dollars above this estimate, according to other presentations by the USDA/APHIS.[23,25]

Prior to the eradication program beginning in 1995, the Department made no regular inspections for citrus canker within residential neighborhoods, with the exception of limited inspections on the west coast of Florida in Manatee County during the eradication program in the 1980's. Without inspections, it is not possible to know the first arrival of citrus canker in Miami-Dade County occurred or for that matter, South Florida, prior to the discovery in 1995. To inspect the entire Miami-Dade County would require enormous resources with 2,431 square miles and approximately 2.6 million residents (2013 estimate).

It has been suggested the introduction of canker might have occurred in 1992 based on early surveys which discovered infected trees with an estimated 2 to 3 year old lesions.[9,13] However, as stated by Dr. Schubert, "The life of a citrus leaf is usually three years." (International Citrus Canker Research Workshop (ICCRW), June 21, 2000, transcript, page 379). There are typically over 70 significant rainstorms in Miami, so there are many opportunities to disseminate the bacteria within the canopy and to surrounding trees every year. Thus, the disease can persist in an area for many years. Scientists have also suggested the discovery of canker in 1997 on the west coast of Florida may have been a holdover from Canker War II (1986 to 1992), based on genetic testing.[21,26]

Detection and proper identification are serious problems. Natural leaf drop can hinder detection. Pruning of diseased limbs and foliage would result in elimination of the older and more developed lesions. The first line of defense against pests and diseases in South Florida is often the pruning shears. Of course, owners of a tree with lesions may simply cut down their tree.

The slow development of symptoms and lack of decline in health can be attributed to the fact the disease is non-systemic, meaning bacteria do not enter the vascular system of the plant.[17] The time to discover citrus canker is not simply a function of latency time, particularly in residential neighborhoods where residents are unlikely to conduct thorough inspections of their trees. The aggregate time delay or "observational lag" includes delays as a result of latency, incomplete and irregular inspections, environment factors (particularly cold weather), natural defoliation, pruning activities and the inability to differentiate symptoms from other foliar diseases at early times.

By year 2004, the FDACS reported that canker had been found in 18 counties, and approximately 1500 square miles were under quarantine. Miami-Dade and Broward counties in South Florida accounted for most of the quarantined area. In May 2005, citrus canker was discovered in a nursery in Polk County.[8] This was the first time nursery plants had been destroyed and nurseries quarantined. This likely made all citrus growers concerned, because if the trend continued, it would likely reduce the available stock needed for their resets (new plantings).

Residents and Community Leaders Oppose the 1900-ft Rule

There was opposition to the cutting program during the 125-ft policy, but this was mild compared with the opposition to the program with the 1900-ft circles, encompassing 260 acres. It was difficult for residents to know whom they should address their concerns as the CCEP was being promoted as a collaborative effort by federal and state agencies.

Citrus canker can be misidentified[6], so both underestimation and overestimation of the number of trees infected are possible. When there is a single misidentification, with the 1900-ft policy, there may be approximately 800 "exposed" citrus trees unnecessarily destroyed. A tree density of 2,000 citrus trees/square mile or 3.12 citrus trees/acre used in this calculation is based on the Florida field study statistics.[16] As stated in an article by Schubert et al.[26], there can be up to 10 citrus trees/acre, thus a single misidentification could result in up to 2,600 residential citrus trees being destroyed.

The Miami Herald, on November 5, 2000 indicated that there were 51 known cases where the Department incorrectly identified other trees as citrus trees based on a review of the CCEP hotline logs. These trees included mango, cherry and avocado. The cherry trees mentioned in the article are Surinam cherry (*Eugenia uniflora*), a common hedge plant in South Florida.

Other complaints cited by the Miami Herald were for damaged sprinkler heads, irrigation lines, and fences. Most likely, dragging the stump grinder equipment through the backyards of residents was responsible. Orchids and other plants were damaged or destroyed as a result of eradications. In one case, a gas pipeline was cut. Fortunately, no one was hurt.

The Department also acknowledged that it was aware of the wrongful destruction of 41 canker-free citrus trees. It was very difficult for a resident to have any evidence of wrongful destruction, due either to misidentification of the disease or type of tree, based on a handful of dust that was left behind after eradication.

Through its public relations efforts, the Department tried to minimize the efforts of program opponents. They frequently informed the public that the disease was both highly contagious and devastating to citrus trees. Also, the Department's public relations officers attacked residents opposed to the program as selfish "tree huggers" more concerned with their few trees than saving Florida from a terrible disease.

Elected representatives in Miami-Dade and Broward counties began to react to complaints of residents. Many mayors and county commissioners were active in approving resolutions seeking legal challenges to the program including Broward County Commissioner Chairman John Rodstrom and Commissioner Lori Parrish. Legal challenges were made to the 1900-ft policy from 2000 to 2004 as a result of the joint collaboration of Miami-Dade and Broward counties and legal departments from 8 municipalities.

These legal challenges identified numerous potential violations of law, including the right to enter properties without search warrants, inadequate notification to owners of tree destruction, circumvention of proper rule-making procedures, lack of fair and just compensation to homeowners for destroyed healthy trees, faulty laboratory procedures in identification of canker and ambiguity in the measurement of the 1900-ft limit. These challenges would not have been possible without the dedicated support of the Broward and Miami-Dade County attorneys, who fought with the Department on behalf of their residents.

There were also attempts to challenge the validity of the research supporting the 1900-ft distance in court hearings. However, these lawsuits could not secure the release of the raw data from the 1998 Florida field study. The Department insisted this research provided strong support for the 1900-ft policy. To date, the raw data and related information including the maps, spreadsheets and computer programs used in the field study have not been made public, despite requests to the USDA under the Freedom of Information Act.

The Florida Supreme Court ruled in 2004 that the canker law could only be challenged under the rational basis standard. To defended the canker law, FDACS needed only to show there was a rational relationship between the eradication program and program's objective of eradicating canker. The rational standard was acceptable to the court, because some compensation for destroyed trees was paid to owners. This ruling eliminated future challenges on the field study research (termed the "Gottwald study" by the courts) leading to the 1900-ft policy.

The courts at various times prohibited the cutting of healthy trees, however they always permitted FDACS to cut infected trees. The courts upheld the right of the Department to remove diseased trees where reliable laboratory identification procedures were followed. Entry onto private property without permission or warrants was ultimately denied by the courts and this added to the difficulties to the CCEP. The results of the major legal cases are summarized in Chapter 5.

After USDA funding was stopped on January 10, 2006, a new program was developed to help control citrus canker, called the Citrus Health Response Plan. The plan focused on keeping citrus canker from entering nurseries, rather than inspecting and eradicating canker in residents' backyards. The new plan is focused on keeping harmful citrus diseases and pests out of commercial areas or liming their dissemination.

During the eradication program, citrus canker was discovered in both the Bahamas[1] and Australia. Neither country adopted the same policies of massive residential eradications as occurred in Florida. The Bahama outbreak was blamed on the shipment of seedlings for planting in the groves from Florida.[1] In both these cases, the outbreak occurred in their commercial areas and was effectively controlled without a full blown 1900-ft policy in both residential and commercial areas.

4. The Elusive Disease

The warm and wet climate of South Florida is very conducive to the dissemination of citrus canker. By January 2006, citrus canker had been discovered in 24 counties of Florida. Citrus canker is an elusive disease and first impressions made at discovery may be wrong. When canker is discovered, the lesions on the tree may be more than two years old. As Dr. Schubert stated at the 2000 International Citrus Canker Research Workshop, "We are looking at history, when we find the disease ..." (June 20, 2000 transcript, page 144 as posted on the website). Citrus canker is known to "overwinter" which means citrus trees infected during the wet summer months may not show identifiable symptoms until the following spring or summer.[21] Dr. Timmer stated at the conference in year 2000:

> "... you have to consider that these things *[biological process by which canker spreads]* can remain dormant on citrus trees for up to two years, which is influencing all this stuff *[canker detection]*"

(June 20, 2000 transcript, pages 319- 320. copy of transcript provided in the website).

The brown and yellow halo around the lesion may diminish or disappear with time so older lesions may be even more difficult to identify as citrus canker.[37] Within residential areas without any formal inspections, the time to discovery can be very long, in terms of years, since residents are not necessarily aware of citrus canker symptoms.

Another misimpression made at discovery is that the source of canker must be from another tree within the immediate vicinity. The initial infection may have occurred in the nursery, the commercial outlet, or anywhere along the supply chain providing citrus for sale. The term "spread" commonly used in plant pathology, is avoided in this book as this term may imply movement of disease outward from initial source of disease by natural forces. Because citrus canker is an elusive disease with multiple pathways, a more general term, "to be disseminate" is used in this book which does not imply an immediate or local source.

Maps showing infected tree locations in groves have been described as patchy or scattered patterns.[34] A highly contagious disease, if the disease is easily identifiable, should result in a clustered pattern or simply blanket an identifiable area. The observed scattered pattern caused a plant pathologist to question the highly contagious aspect of citrus canker.[34]

5. Pieces of the Puzzle

When the 1900-ft policy was announced in a news release on February 11, 2000, the only information available to the general public was a single page brochure from Commissioner Crawford, thanking residents for their cooperation as the Department takes the necessary action to eliminate a devastating disease. The lack of supporting research on the 1900-ft policy seemed very odd. Public relations officers with the Department stated at public meetings that a one-year field study had been conducted by the USDA. Yet, residents who searched the internet could find no information on this field study from either the Department or USDA website. All requests from the USDA for information were denied. Residents were told that a report was being prepared for publication in a scientific journal at an unspecified later date.

In February 1998, the Department initiated a moratorium on healthy tree cutting in Miami-Dade and Broward counties and authorized a one year field study with the goal of tracking canker. So, it might be assumed that by March or April 1999, this study would be complete and at least some basic information would be available to the public. This was not the case. No basic information was available to the public prior to year 2001. This information would include the survey forms used to collected data, dates when the inspections occurred, boundaries of the sites and the prevalence of citrus trees within the site. Probably the interesting data would be the initial conditions of the sites or exactly how many trees were healthy on the first survey. Maps showing the locations of discovered infected trees in sites would also be basic information. There was nothing provided by the Department. In fact, until today, none of this basic information has been released by the Department or the USDA.

It was also quite surprising to learn that a retired senior scientist with the University of Florida/IFAS, Dr. Whiteside with decades of experience in citrus diseases, testified in the court in November 2000 that it was impossible to track citrus canker as it went from tree to tree. His testimony centered on how often inspectors miss the symptoms of citrus canker when inspecting commercial groves.

The detection problem had been discussed at the 2000 International Citrus Canker Research Workshop for grove surveys. According to Dr. Armando Bergamin-Filho increasing the number of inspectors in commercial groves, consistently resulted in more discoveries of canker. The detection problem is compounded by false positives, or the misidentification of other diseases as citrus canker, as discussed in the workshop. The problems of misidentification of citrus canker are legendary as the Department spent nearly five years destroying millions of nursery plants for "false canker." Inspections for citrus canker, should properly be termed a search, identify and test mission.

Previous articles suggested the disease could be disseminated from a single tree to others up to 7 to 8 miles away.[9,19] However, associations between infected trees in residential areas, which had never been thoroughly inspected before, seems highly speculative. To identify movement this large, it would appear that one would need a remote island, where every citrus tree had somehow been certified as canker-free, except for one disease tree.

Still another puzzling bit of information surfaced in November 2000 during a legal challenge to the program. Dr. Gottwald of the USDA/Agriculture Research Service testified in Broward Court, that a meeting had been held in December 1998, in which a consensus of attendees agreed on the 1900-ft rule. This was consistent with the January 2001 article in Phytopathology on the field study.[14] Yet, for this important meeting, the Department could produce no list of attendees, meeting agenda, minutes of the meeting or copies of Dr. Gottwald's presentation. Was this key meeting conducted in secret? Did it actually happen? Not a single news story appeared on this meeting.

The Department seemed to have difficulty answering simple questions, particularly concerning the field study's methodology and data collection. Where were the study sites? When did the data collection begin and end? Scientists were supposed to convene every month to review the field study. There were no minutes of these meetings, at least, nothing made public. In response to a query on field study information, the Department stated they did not retain any of the collected survey sheets or maps showing the locations of infected trees within the study area. This seemed very suspicious. Certainly, there was information such as the intensity or severity of the disease in each of the sites, which would have been of interest to the Department. Why would all the unique data collected for over a year by Department inspectors not be retained by the Department at least in electronic format? All this seemed very strange, and certainly it suggested FDACS had something to hide.

The old saying, "The devil is in the details" holds true for the eradication program. A major piece of the puzzle is specifically how the Department arrived at a 1900-ft radius. Certainly, state and federal officials would want to know how effective any suggested radii would be before committing enormous resources to the eradication program. They would want to know all their options and the impact of any particular radius. One presentation by the field study's primary investigator, Dr. Gottwald, indicated that the 1900-ft policy would capture 99% of the subsequent diseased trees.[5] Yet, the Department on their website and in numerous other presentations stated that the 1900-ft rule would capture 95% of subsequent diseased trees.

How was the science behind the 1900-ft policy reviewed? The Department stated that an epidemiology study had been presented in numerous times prior to year 2000 including a Task Force meeting in May 14, 1999. Yet, the minutes of this meeting provided no details on study, except for a single page of results from one of the sites. For 13 months after the start of the 1900-ft policy, there was nearly no information on the study. Then, in January 2001, an article in the form of a Letter to the Editor was published in Phytopathology, with the impressive title, "The Citrus Canker Epidemic in Florida: The Scientific Basis of Regulatory Eradication Policy for an Invasive Species." This article provided only a few details, as the Department stated in personal correspondence, "... the letter does not have a full description of methodology, results, and discussion of analysis." In fact, there was no description of the methodology in the 5-page article.

A second article was published in April 2002, entitled, "Geo-Referenced Spatiotemporal Analysis of the Urban Citrus Canker Epidemic in Florida." Critical basic information seemed to be missing from this article as well. It was published well over two years after the 1900-ft policy began, and likely at least a year after it was anticipated that residential cutting would be completed, and yet did not provide any real recommended distance for complete eradication of canker.

When at last some field study information was revealed, there was an abundance of conflicting information. Numbers just seemed to be dancing around. An USDA project status report ending June 1999, indicated more than 12,000 trees were surveyed.[10] An initial report[11] stated that there were more than 13,000 trees in the study, yet in another report based on May 11, 1999 presentation at the Citrus Canker Risk Assessment Group meeting (CCRAG) by the same authors, stated <u>nearly</u> 15,000 trees were surveyed.[5] At that point, it looked like the number had to be between 13,000 and 15,000 trees. However, in a published article, in January 2001, the number of surveyed trees had grown to approximately 19,000 trees.[14]

A mishmash of conflicting "distances of spread" values was made public. A January 2001 article published in Phytopathology identified a maximum distance of 11.1 miles from one tree to the next.[14] Prior presentations at meetings were consistent with this maximum value. Yet, in another article, published in the same journal (Phytopathology) by nearly the same group of authors in April 2002, the maximum distance was 2.2 miles[16] and a whole new set of distance values was presented without any explanation. This seemed very strange.

The Department was constantly alerting residence that citrus canker is highly contagious and "spreads like a wild fire." Yet in one study site, Site D3, only about 3.2% of the trees became infected over the course of the study and yet in another site (Site D1) 29% of the trees were infected.[16] Which set of results was most representative of the disease progression?

The published articles on the field study[14,16] made no explicit recommendation on cutting radius distance. Nor was there any evidence introduced in 2000 Broward Court hearing showing any recommended distance based exclusively on USDA research. The origins of the 1900-ft rule seemed to be an open question.

There was also the question of how citrus canker became disseminated on both the east and west coast of Florida and in geographically remote areas of Florida. How did citrus canker become disseminated from the original discovery in Miami to Big Pine Key, more than 50 miles to the south by year 2002 and to Polk County, more than 150 miles to the north by year 2005?

This book is the first independent documentation on the field study and other aspects of the program. For now, it is best to set a proper foundation, in the next several chapters, before exploring how the pieces fit together.

6. Summary of Citrus Trees Destroyed and Costs

"If we have to go beyond 125 ft, we are woefully under-funded...," Deputy Commissioner Craig Meyer, Citrus Canker Task Force meeting, May 14, 1999.

Trees Destroyed during 1995 to 2006

A total of 16.5 million trees were destroyed from 1995 to 2006 based the FDACS 2012 Comprehensive Report on Citrus Canker Eradication Program in Florida.[8] A copy of this report is provided on the online supporting documents website. The cutting statistics for the post 2000 period alone, when the 1900-ft policy was in effect, have not been disclosed by the Department.

Table 1.1: Eradicated Residential, Nursery and Commercial Trees

	Citrus Trees Destroyed	Percentage of trees in each category
Residential	850,643	5
Nursery	4,334,154	26
Commercial	11,323,298	69
Total	16,508,095	100

As shown in Table 1.1, of the 16.5 million trees destroyed, approximately 5% were residential trees, and the remaining 95% were either nursery or commercial trees.

Canker was discovered in 24 counties of Florida. However, only four counties (Highlands, Polk, Hendry and De Soto counties) account for all of the 4.3 million nursery trees destroyed as shown in Table 1.2. These counties would typically be considered within Central Florida, where most commercial citrus trees are grown.

The Comprehensive report states that 87,493 acres of commercial groves were destroyed or approximately 129 citrus trees per acre. As a comparison, the density of plantings in urban residential areas is estimated to be approximately 3.13 citrus/acre.[16] South Florida was very different from the rest of Florida in comparison to where canker was found and eradicated. In South Florida, 59.3% of eradications occurred in residents' yards and the remainder, 40.7%, were in commercial areas. Outside of South Florida, 1.2% of the eradications occurred in residents' yards and the remainder, 98.8%, were in commercial areas, based on data shown in Table 1.2.

Table 1.2: 1995 to 2006 Eradicated Trees per 2012 Comprehensive Report[8]

	Commercial Trees	Nursery Trees	Residential Trees
Monroe	0	0	519
Miami-Dade	462,701	0	463,437
Broward	See note*	0	180,875
Palm Beach	28,867	0	72,184
Total - South Florida	491,568	0	717,015
Polk	167,437	1,677,376	788
Highlands	1,935,267	1,871,346	260
De Soto	1,052,381	120,361	241
Hendry	1,052381	655,071	862
Other counties	5,248,287	0	131,477
Total - Other areas	10,831,730	4,334,154	133,628
Total - Florida	11,323,298	4,334,154	850,643

* Miami-Dade and Broward counties were combined for commercial trees.

Total Trees cut = 16,508,095
Total trees cut, South Florida = 1.2 million trees
Total trees cut, All other areas = 15.3 million trees

Quite surprising, De Soto, Highlands and Okeechobee counties had no discoveries of canker within residential areas, although some exposed trees were cut. Possibly, these exposed trees were cut and later lab tests failed to confirm canker on the presumed infected trees. It is also possible the Department destroyed citrus in abandoned lots, which would not require identification of canker.

It should not be concluded that the canker disease was more concentrated in South Florida. South Florida had been surveyed for 10 years, and during the first 4 years of the program, the surveying concentrated on Broward and Miami-Dade counties. Without an estimate of the total number of citrus in these residential areas and the degree to which areas were surveyed, no clear conclusion on the concentration of the disease can be drawn.

Cutting Statistics using 125-ft and 1900-ft Rule

The 2012 Comprehensive Report[8] provides a tally of trees cut during the program, by county and growing environment (nursery, grove and residential) for all years of the program. There is no breakout in these statistics on the number of residential healthy and infected trees destroyed as a result of the 125-ft and 1900-ft rule. Cutting statistics are not given by year. The 125-ft rule was applied both in residential areas and in the lime groves, in the Homestead area of South Florida.

A 1900-ft circle is 232 times larger than a 125-ft circle, so many more healthy trees would be destroyed for every infected tree. Because eradication circles tend to overlap, a calculation of healthy to infected tree ratio (H/I ratio) can not be based strictly on geometry. To plan for eradications, officials would require some estimate of the degree to which the circles overlap and the citrus density (trees per acre or square mile) for residential areas, groves and nurseries.

The USDA/APHIS, Office of Inspector General Audit Report states there were 16 healthy trees for every infected tree cut down (H/I ratio = 16) in the period of November 2000 to February 2002.[29] Using the FDACS 2012 Comprehensive Report[8], the H/I ratio for residential areas can be calculated for counties where canker was discovered after year 2000. In these counties, the average H/I ratio is 30, with a range of 14 to 200. The high end of this range (H/I = 200) is from Indian River County, where 8 infected and 1597 healthy trees were cut. Non-overlapping circles would result in the highest H/I ratios. In summary, it is estimated that between 16 to 30 healthy trees were cut for every infected tree, within residential areas based on both the USDA/APHIS audit[29] and FDACS report[8]. This is the equivalent to 94 to 96% of the trees cut down were healthy trees.

Eradication Forecasts

> Canker fight may wipe out 1 million trees. Emergency bid to save fruit industry may burn a third of S. Florida Citrus.

The Miami-Herald headline of February 12, 2000.

Initial planning likely required forecasts of the trees to be eradicated based on 1900-ft policy or any other chosen radii. Based on a viewgraph presented in November 9, 2000 in Broward Court by Dr. Gottwald, an estimated 750,000 to 1,000,000 trees were to be destroyed. He presented a similar estimate during the June 2000 International Citrus Canker Research Workshop. This is very close to the actual value of 717,015 residential trees cut in South Florida or 850,643 residential trees cut in Florida. Concerns about the legality of the healthy tree cutting had been expressed by both FDACS Deputy Commissioner Craig Meyer and Mr. Richard Gaskalla, Director of the Department of Plant Industry during the Task Force meetings. The healthy tree cutting forecast would also help the Department determine their liability in case of lawsuits. This eradication forecast is an important part of the origins of the 1900-ft rule as explained in Chapter 8. However, this is jumping ahead in our story.

Program Costs

The operational cost is estimated to be 521 million dollars as shown in Table 1.3. The CCEP costs are based on a UF/IFAS study, available on the UF/IFAS website.[35] Eradication costs prior to year 2000, under the 125-ft rule, total 66 million dollars, while the costs for years 2000 forward were 455 million dollars. Had the program ended in year 2000, after the Department had been ordered by the Broward civil court to cease healthy tree cutting, the eradication costs would have been 194 million dollars.

Table 1.3: Approximate Operational Costs of the Eradication Program

Year	125-ft Operational cost ($1,000)	Year	1900-ft Operational cost ($1,000)
1996	3,080	2000	128,219
1997	6,808	2001	94,540
1998	13,836	2002	79,856
1999	42,596	2003	68,699
		2004	38,718
		2005 (Estimate)	44,000
		Total	521,352

There is almost an annual doubling of eradication costs, going from 3 million in 1995 to 6.8 million in 1996, then 13.8 million in 1997. By 1997, canker had been found in many locations in Broward. This raises the question if the increase in discoveries was because there were more inspectors, looking over a larger area or was canker actually spreading? Since there were no prior inspections of residential areas, there is no way to know which is true. There is a significant decline in operational costs after year 2003. After 2003, the focus of the Department seemed to be more on groves than residential areas. Both inspections and eradications also would be less expensive in groves than in residential areas. These are operational costs, and do not include compensation to commercial groves and nurseries.

The USDA spent approximately $636 million for producer compensation according to the Lakeland Ledger article of October 6, 2006. This estimate included claims which were still being processed. Thus, the sum of the compensation and operational costs is approximately 1.2 billion dollars.

The $636 million in producer compensation is likely underestimated. More recent documents from the USDA/APHIS/PPQ indicate the total program cost to be more than 1.3 billion dollars[23,25], although these are not official reports. The Palm Beach Post indicated on March 19, 2015 that the cost could be as high as 1.6 billion dollars. It is likely these higher estimates include payments to producers as part of the Risk Management Agency's (RMA) Florida Fruit Tree Pilot Crop Insurance Program. The pilot program begun in year 1996, with the purpose to cover damages and destruction due to freezes, windstorms and other weather related events. It was extended in year 2000 to include the destruction of positive and exposed trees under the CCEP. Thus, the compensation cost of 636 million dollars may not include crop insurance program payments, perhaps an additional 300 million dollars for the six years.

Legal expenses have been excluded from this total, as there is no accurate accounting to date. FDACS has relied on outside legal firms. The study shows no legal expenditures, except 5.2 million dollars in year 2001. The Department has refused to pay for judgments from the class action lawsuits which may total up to 100 million dollars.

The estimated compensation costs of 636 million dollars or roughly 50% of the program cost was for producers' losses. The USDA stated in year 2000 (Payments for Commercial Tree Replacements as published in the Federal Register),

> We consider that trees infected with or exposed to citrus canker, because of the destructive nature of the disease, have no value. Thus, the tree replacement payments provided for by this interim rule are intended to provide eligible growers with the funds necessary to establish new plantings, rather than to pay for the trees destroyed because of citrus canker.

Yet, in 2011, the Office of Inspector General for APHIS, investigated five large growers and none of them had replanted citrus.[31]

Did the compensation program for grove owners actually promote the decline in Florida's citrus industry, by motivating grove owners to destroy their groves, and sell their land at a large profit? A single infected tree at the center of 1900-ft circle (260 acres), would result in destruction of between 26,000 to 39,000 healthy trees, given tree spacing is 100 to 150 trees/acre. It would be very tempting for citrus growers to find an infected tree, and move it to the edge of the field, to benefit from an overly generous "replanting" program.

Payments to Grove Owners to Support Recovery

The Office of Inspector General (OIG) made a survey of groves receiving compensation and found none of them used the funds to replant citrus:[31]

> Our onsite reviews of four growers who received compensation as a result of the 2005 hurricanes in Florida found that the growers had sold the land or otherwise not replaced citrus trees more than 2 years after receiving their payments and 3 years after the trees were destroyed.
>
> - Grower A had not replaced any citrus trees on 2,887 acres, but instead sold the acres for $20.2 million even though he had received $8.82 million in citrus canker tree replacement payments for the same acreage.
> - Grower B had not replanted the 4,550 acres where citrus trees were destroyed, even though he had received $14.25 million in citrus canker tree replacement payments.
> - Grower C had not replanted the 3,311 acres where citrus trees were destroyed, even though he had received $8.83 million in citrus canker tree replacement payments.
> - Grower D had not replanted more than 270 acres where citrus trees were destroyed after receiving $730,000 in citrus canker tree replacement payments.

In addition, OIG showed APHIS had overpaid grove owners by approximately half a million dollars and was not interested in trying to have the funds returned.[31]

The Contributions of Dr. Jack Whiteside

The Department contended there was almost total agreement to its policies among researchers who understood citrus canker the best. However, opponents to the 1900-ft program soon learned this was not true. Plant pathologists either supported the program, or stayed on the side lines, not wishing to cross lines with the Department.

However, Dr. Whiteside was not one to keep to himself. From his extensive experience in Zimbabwe and Florida, he had learned that the disease was not devastating, and the option of living with canker, and taking preventative control measures were viable and perhaps less costly options.

His candid comments of how inspectors can often miss symptoms of the disease because they are too small, located high in the new flushes of a tree or not easily distinguishable from other diseases is the hard reality of canker surveys. The elusive nature frustrates eradication and the study of infected tree patterns created by the epidemic. Residential areas have even more problems, including the actions of homeowners who likely prune infected limbs and remove infected fruit, making detection difficult or impossible.

His viewpoint was that it is extremely rare that a plant disease can be completely eradicated, unless conditions for its survival are marginal. Additionally, he strongly supported managing the dissemination of the disease through windbreaks and copper spraying in the groves. References 32, 33 and 34, at the end of this book cite his articles from 1985 to 1988 and are posted on the website.

It is unfortunate that the Department decided to criticize this work, by posting an unfair critique of his three papers to their website, without posting his articles. In many ways, Dr. Whiteside's articles were actually ahead of their time, with his emphasis on windbreaks and copper sprays. Today, most growers would also include the control of citrus leafminer and high levels of disinfection and sanitation in the groves as essential in controlling canker. Dr. Whiteside would have been pleased, as change is always better late than never.

7. Participating Organizations

Table 1.4 provides the name and abbreviations of the participating organizations. All primary participants in the program were from state or federal government agencies, and the University of Florida institute facility in Lake Alfred, Florida. Professors from the University of Florida main campus in Gainesville, FL were not involved in the program.

Table 1.4: Organizations Involved in Citrus Canker Eradication

Agency	Full Name	Abbreviation
State Governmental Agencies	Florida Department of Agriculture and Consumer Services	FDACS
	Division of Plant Industries	FDACS/DPI
Federal Government Agencies	United States Department of Agriculture/Agricultural Research Services	USDA/ARS
	USDA/Animal and Plant Health Inspection Service	USDA/APHIS
	USDA/APHIS/Plant Protection and Quarantine	USDA/APHIS/PPQ
	USDA/ Plant Health Board	
University	University of Florida/Institute of Food and Agricultural Science	UF/IFAS

Who ran the CCEP?

Since this was a joint cost sharing program between the State of Florida and the federal government organizations, the overall responsibility of the program was shared at the very top level between the Governor of Florida and US Secretary of Agriculture.

At a more detailed level with authority, the Commissioner of FDACS would direct and approve actions of the CCEP. The program would have to be within the budget as approved by the legislature. The Commissioner of FDACS is an elected position. From 1961 to 1991, the Commissioner was Doyle Conner, succeeded by Bob Crawford from January 1991 to January 2001. Charles Bronson was Commissioner from January 2001 to January 2011.

Directly below the Commissioner during Canker War III, was Deputy Commissioner Craig Meyer, an appointed position. As Deputy Commissioner of Agriculture, he was the top official with the Agricultural Department and would have primary responsibility for implementation of the 1900-ft eradication policy, coordination with the USDA, program funding, legal issues and public relations. The Deputy Commissioner and his staff, were located in Tallahassee, Florida. Craig Meyer testified in the Broward and Miami-Dade court cases. Mr. Meyer is an attorney of law.

No one person was designated as the "CCEP Project Director" an individual who would be exclusively devoted to all aspects of the program and have no other responsibilities. Perhaps this was because the 1900-ft program was ostensibly considered a one-year program. The individual who seemed to take on the most responsibility was Richard Gaskalla, as the Director of Division of Plant Industry of the FDACS. He acted as the central administrator or coordinator of the program, in addition to his normal duties.

For residential eradications in year 2000, Ken Bailey, FDACS/DPI, Director of CCEP within Miami-Dade and Broward counties, was in charge of the operational aspects of residential program. His responsibilities would include leasing offices, hiring inspectors, developing inspection protocols and training and signing removal contracts with tree maintenance companies. Michael Hornyak, USDA/APHIS/PPQ, Miami-Dade and Broward counties, monitored the progress of the program for both residential and grove eradications for the USDA and coordinated USDA inspections with the CCEP.

Mr. Leon Hebb was in charge of the Bureau of Pest Control and Eradication, in Winter Haven, Florida, and was a Co-Director of the CCEP. He reported to Mr. Richard Gaskalla and was responsible for the commercial eradication program. He was a key participant in the Task Force which recommended the 1900-ft rule.

Dr. Stephen Poe, as Program Coordinator for the USDA/APHIS/PPQ, was responsible for compensation for growers, and quarantine rules. He was also a key participant in the Task Force. Mr. Mike Shannon, State Plant Heath Director, USDA/APHIS/PPQ worked jointly with Dr. Poe in coordinating the role of APHIS in the program.

The locations of the various groups were spread out across Florida. The distance between Tallahassee and Miami is approximately 500 miles, so air travel was likely needed for participants of the CCEP to meet in person. The location of each organization is provided on the online supporting documents web site.

Key Personnel in Research and Technical Guidance of the CCEP

The positions and professional affiliations are given below as of year 2000:

Dr. Timothy Gottwald, Research Leader and Plant Pathologist at the USDA, Agricultural Research Service (ARS) in Ft. Pierce, Florida. He is a highly regarded plant pathologist and epidemiologist, specializing in citrus diseases with an extensive list of publications. Dr. Gottwald was an integral part of the discussion that preceded the implementation of the 1900-ft rule. He was called as an expert witness for the Department in two court hearings in Broward County. He was the principal investigator in the Florida field study.

Dr. James Graham, Soil microbiologist and Professor at the University of Florida/IFAS. He participated in the epidemiology study and co-authored many articles with Dr. Gottwald including two key articles on the Florida field study.

Dr. Wayne Dixon, Bureau Chief of the Entomology, Nematology and Plant Pathology Section at FDACS/DPI. He was a key participant in the CCEP decision making as group leader of the Citrus Canker Risk Assessment Group.

Dr. Timothy Schubert, Administrator for the Plant Pathology Section at FDACS/DPI. He has decades of experience in citrus canker. He was part of the risk assessment group and attended meetings of the Task Force (non-voting). He was an expert witness for the Department in the Broward court cases, testifying on the necessity of the CCEP.

Dr. Xiaoan Sun, Plant Pathologist at FDACS/DPI. He was a participant in the Florida field study in 1998, as he was responsible for identifying the oldest lesion ages on the trees in the study sites. He also authored an article on the "Wellington" or A_W strain of ACC as discussed in Chapter 3.

Dr. John W. Miller, Plant Pathologist at FDACS/DPI. Dr. John Miller was a member of the Risk Assessment Group.

In addition, Dr. Gareth Hughes, Professor at the University of Edinburgh, Scotland, specializing in plant disease epidemiology, was a co-author of the January 2001 Letter to the Editor in Phytopathology, which provided results on the field study. Dr. Frank Ferrandino, Scientist with the Department of Plant Pathology and Ecology, in New Haven, Connecticut was a co-author of the April 2002 article on the field study in Phytopathology. Neither of these two researchers appeared at the task force meetings, public hearings, public meetings or court venues to explain the research. Their direct involvement in the 1998 field study in Florida was likely quite limited.

The online supporting documents website provides additional biographical information on Drs. Gottwald and Graham, and the organizations involved in eradication program and relevant research.

8. Moving to the 1900-ft Policy

The eradication program presumably considered elimination or near elimination of canker as a much more preferable option than attempting to control the dissemination of the disease. Dr. Whiteside and others considered disease management as the best economical option, based on decades of experience.

The policy began based on one core assumption — a larger radius would be more efficient than a smaller one. However, in selecting how large a radius to use, many other factors had to be considered such as legal issues, reactions of residents, logistics and the added cost of destroying many healthy trees. Also, policy makers had to decide if the same eradication policy was to be adopted for both commercial groves and homeowners.

Two groups were instrumental in advising regulators for a change in policy, the Citrus Canker Risk Assessment Group (CCRAG) and the Citrus Canker Technical Advisory Task Force (CCTATF). Although the Risk Assessment Group and the Task Force made recommendations to the CCEP, these groups were not established by law. The Department was not obligated to follow their recommendations.

Citrus Canker Risk Assessment Group

The Citrus Canker Risk Assessment Group (CCRAG or RA Group) consisted of 7 plant pathologists from the USDA, FDACS and UF/IFAS. The RA Group was headed by Dr. Wayne Dixon, FDACS/DPI and the members in 1999 were Dr. Stephen Poe, USDA/APHIS, Dr. Timothy Gottwald, USDA/ARS, Dr. James Graham, UF/IFAS, Mr. Leon Hebb, Dr. Timothy Schubert, and Dr. Xiaoan Sun. The latter three members were with the Department. Dr. John W. Miller, FDACS/DPI, was a member in 1998, and likely replaced by Dr. Sun during 1999.

The RA reports from 1998 to 1999 included: (1) Adjustments to the 125-ft policy for grove owners on a case by case review, (2) Recommendations related to movement of citrus fruit from groves found with citrus canker, (3) Adjustment of quarantine limits for groves and (4) Sanitation and processing regulations related to shipment of fruit from quarantine areas. Generally, the exceptions to policy came only after a review of the discoveries, and were contingent on access to properties for resurveys and good sanitation practices within the groves.

After the 1900-ft policy was implemented in year 2000, the RA Group likely continued their work of recommending adjustments to the eradication policy for commercial groves on a case by case basis until Administrative Judge Laningham ruled on July 31, 2001 against the risk assessment program. He found the list of risk factors in the Department's rules lacked any real criteria for evaluation, thus were too undefined, allowing for subjective judgements.

In general, the RA reports can characterized as "grove owner" friendly in their recommendations concerning the 125-ft policy for west coast grove owners, provided the owners were compliant with the sanitation and survey needs of the program. For example, the RA Group recommended on June 17, 1998 to allow the citrus groves owned by William Grimes and Gary Guthrie not to destroy any healthy trees, after citrus canker was discovered in their groves. Their groves would be considered the positive blocks as "experimental plots" to be resurveyed regularly and observed. Later, when positive trees were discovered on two more blocks in the Grimes grove, the Group recommended on February 15, 1999, the removal two trees in each direction from the infected tree. If this was two trees within a row of trees, this would be limited to four trees. The group approved the movement of fruit from positive blocks for juice processing, once eradication was complete.

The RA Group was far less friendly to eradication policy for residential properties. There was no advocate for the residents at these meetings. Also, it was impractical to perform risk assessments on a case by case basis. For the Miami-Dade County program, the RA Group concluded on March 4, 1999 that the collection of samples for a second opinion was deemed unnecessary and "posed an biological risk of spreading canker" as this constituted movement from a quarantined area to a laboratory located in a non-quarantined area. Further, the group stated, "The policies and procedures used by the department to diagnosis citrus canker are scientifically proper and accurate." The reliability of laboratory testing would be challenged in Broward Court. On July 18, 2003, Judge Fleet ruled the testing for citrus canker was inaccurate. The decision was later reversed by the District Court of Appeals on a technical basis of the appropriate venue to hear the case.

The RA meeting of May 11, 1999 was critical step forward in the enactment of the 1900-ft rule. This meeting included was attended by 11 other invited guests. Dr. Gottwald made a presentation based on the "epidemiology study" of which only sparse, selective information on the study was presented in the RA report. The RA Group recommended 1900-ft for the Miami-Dade and Broward counties eradication program. Cutting should continue with a 125-ft circles until resources are available to eradicate with 1900-ft circles. A table of results for the epidemiology study was attached to the RA report. A footnote at the bottom states that the table was revised on May 26, 1999. Further, the only reporter present at the meeting, Mr. Paul Power of the Lakeland Ledger did not report any recommended changes in eradication policy. Therefore, it is not clear if these recommendations were made at the meeting attended by Mr. Power. The inclusion of a recommendation in their report may have been added later, with the agreement of the members.

The RA Group produced a report entitled, "Bacterial Citrus Canker and the Commercial Movement of Florida Citrus Fruit" on July 14, 1999. The report as posted on the FDACS website appears to be only a draft copy.[36] The primary conclusion is that properly sanitized and processed citrus in areas where canker is present, will have an extremely low chance of spreading the disease. Other conclusions were that it is extremely difficult to track disease movement from source to newly infected tree.

The recommendations or approvals made by the RA Group or Task Force were not binding on the Department nor the USDA. It is likely however, the opinions provided by these groups were critical in the legal defense of the program. The Department submitted the May 11, 1999 RA report to the Broward Court after Judge Fleet ruled the Department had to release all information relevant to the epidemiology study.

February 1998: Moratorium and Florida Field Study

In February 26. 1998, the Commissioner Crawford, declared a one year moratorium on healthy tree cutting, and proposed a citrus canker study be conducted. The moratorium might seem odd, given that citrus canker had been discovered in Broward and Manatee County, thus greatly expanding the threat of canker. In the press release, the Commissioner briefly identified the study plan:

> The establishment of experimental research zones- subject to the approval of the U.S. Department of Agriculture- to track the spread of the disease from infected tree to exposed one.

> Scientists in the program will designate properties in highly-infected areas, moderately infected areas and low infected areas to determine the rate at which exposed trees are being infected. While scientists will review their findings each month, it is envisioned that the experiment will go on for a full year.

Select topics in epidemiology as related to citrus canker are discussed in Chapter 6. The technical review of the field study is presented in the appendices, available on the online support website. The study conducted by Dr. Gottwald as the primary investigator in five sites in South Florida is termed the "Florida Field Study" in this book to avoid any ambiguity to other research.

The moratorium ended on June 17, 1999, and the 125-ft policy resumed in Miami-Dade county. This continued until year 2000, when the 1900-ft policy was promulgated by Commissioner Crawford.

When did the field study actually begin? One would naturally expect the study to begin in March 1998, after it was authorized by Commissioner Crawford. The published articles on the field study[14,16] do not mention a clear start and end dates to the study. The 2002 article states that there were many queries on the inadequacies of the 125-ft policy and that a "cooperative CCEP, ARS and University of Florida research effort was established in August 1998" to respond to these queries. Was there really a 5 to 6 month lag between the Commissioner's announcement and the start of the field study? Appendix A further explores questions on the field study as provided in the online website.

May 1998: USDA Science Review

The USDA/APHIS convened a Science Review Panel to examine and comment on the citrus canker eradication program in early 1998. A memo from Dr. Poe, of USDA/APHIS/PPQ, to panel members with comments on the panel's recommendations is dated May 13, 1998. The exact dates when the Panel met is unknown. The Panel consisted of a number of university plant pathologists, and headed by Dr. C. Lee Campbell, a highly respected authority in plant disease epidemiology. Dr. Campbell is well known author of a graduate level university textbook on plant disease epidemiology. One would think that a review with outside experts would be the perfect opportunity to share with them the planned field study including the data collection methods, the study's duration and location, and statistical methodology. However, there is no record that any details of the field study were shared with the Panel during the review.

The program of cutting exposed trees was shut down when the panel met so the scope of the "Science Review" was limited to interim measures to control canker, until an epidemiology study was completed. At least one meeting was held among the researchers, but no minutes of the meeting have been made public.

The Panel produced a brief six page summary of their findings. The summary stated that there should be declining incidences of the disease with distance from the origin of the disease (initially infected tree). Also, they state at present that there is no reliable information to determine the distances that the bacteria might spread under urban conditions. They note that the ultimate spread of the disease might be more than a mile from a point of origin.

In their recommendations, they seem to support control measures, as they state, "regulatory personnel should discuss with tree owners strategies for keeping trees disease free." Also, they state that as an interim measure, trees should be removed at a distance of twice the height of an infected tree. For a 30-ft tree, this would mean that a 60-ft radius should be used.

Since the field study would be completed by mid-1999, it would have seemed reasonable to make a second USDA review at that point. However, no further review was ever done, and it is possible researchers did not want to discuss the possible flaws in the field study.

December 1998 Meeting on Canker Eradication Program

According to Dr. Gottwald testimony in the Broward Court on November 2000, the key decision makers had reached an agreement to proceed with a 1900-ft rule in December 1998 at a meeting at the USDA/ARS offices in Orlando, Florida. Dr. Gottwald testified:

> No, the report does not discuss the meeting that occurred in December 1998 in which there was a group of scientists, regulatory agents and growers present in a room and decided upon 1900 feet. 1900 foot [rule] is not decided upon this, the manuscript or anything else, it was decided upon by a group, a body of regulators, university scientists as well as ARS scientists, etc. It was not decided upon by these reports.

According to testimony in the Broward Court, there was no known minutes of this meeting, no prior public announcement of this meeting and no list of attendees. No newspaper ever reported this meeting. Yet the Department went on the record that it occurred as described by Dr. Gottwald. Why would Dr. Gottwald and the other scientists involved in the field study attempt to present results so early? The December 1998 meeting would become one of the "pieces of the puzzle" which will be revisited in Chapter 9

Compensation for Homeowners

A short memorandum prepared by Ms. Connie Riherd, Assistant Director of FDACS/DPI on March 1, 1999, estimated a $365 to $468 the value of an "average" residential tree. The value of $468 was calculated based a formula from the Southern Urban Forestry Association. A handwritten note on this email indicates a lower value is possible ($365/tree) using a different formula from the ISA.[4] However, these evaluations were not known to residents in year 2000. Nor, did residents know that Dr. Gottwald had forecasted up to a million residential trees would be cut as a result of the 1900-ft rule.

Examination of the Environmental Effects

Residential trees can not be cut down in Miami-Dade without a permit. There are numerous exceptions, but only a few would apply to homeowners. These rules are in effect to safeguard against canopy loss and destruction of the natural beauty of communities. Opponents of the program were interested in efforts undertaken by the FDACS in the environmental impact of the expected destruction of one million trees, as forecasted by the Department in cooperation with the USDA. The answer was simple — none.

The USDA/APHIS/PPQ complied with legal requirements by preparing the March 1999 Environmental Assessment[28] , which is provided in the online supporting documents website. At this time, the 1900-ft rule had not been formally adopted, so the assessment did not consider this policy. Nor was there any mention of the eradication forecast. Of course, the USDA could have amended the assessment after the 1900-ft rule was announced, based on the estimated level of destruction to the canopy in residential areas. The USDA commented (Dr. Stephen Poe, Operations Officer, USDA/APHIS/PPQ) that this was unnecessary, as the rule did not significantly change the assessment.

All citrus species are host plants to the giant yellow swallowtail, one of the most beautiful butterflies in Florida. Also, all citrus species are host plants to the Schaus' swallowtail, an endangered butterfly, known only to inhabit the Florida Keys. While canker was found in the Big Pine Island in the Florida Keys and eradication efforts were undertaken, eradication was very limited.

The host plants are those plants which the butterflies lay their eggs, and subsequently their larvae (caterpillars) will consume their foliage. The giant yellow swallowtail must locate citrus plant to lay her eggs. They can not breed in the groves, as they are regularly sprayed with insecticide. So, residential trees are needed for their survival. Additional information is provided in the online supporting documents website.

9. The Citrus Canker Technical Advisory Task Force

On February 12, 1999, Commissioner Bob Crawford announced the formation of the Citrus Canker Technical Advisory Task Force or CCTATF. This is referred to as the Task Force for convenience. This was approximately one year following the announcement of the moratorium and Florida field study. Per the Commissioner's Press Release, "Commissioner Crawford said the task force will be asked to come up with recommendations on procedures that can be put into place to rid Florida of canker and on the critical question of how best to fund those plans."

If general acceptance of the 1900-ft rule had occurred in December 1998 with the principal stakeholders, why review it a second time? In general, it would appear the citrus industry associations had not accepted the 1900-ft rule and it was likely the Department was concerned about lawsuits. The acceptance by the Task Force would not occur until a meeting on July 16, 1999. The recommendation for the 1900-ft rule would be contingent on risk assessment done prior to the eradication.

The Task Force was likely formed for other objectives, including:

- To help coordinate the eradication effort and keep the citrus industry representatives informed on the legal, funding and regulatory issues. Regulatory issues included movement of plants and fruit in quarantine areas of the groves and sanitation regulations Minutes of the Science Working Group and Regulatory Working Group are posted on the online supporting documents website.

- To have on the record, a leading scientist with the USDA/ARS, Dr. Gottwald, recommending to the Task Force the 1900-ft based results of a field study. As stated in the minutes, Deputy Commissioner Meyer stated that the 1900-ft policy might be challenged in court, after the removals had been done, and wanted the scientific rationale for the policy established through presentations at the Task Force meetings. Later, the public relations officers with the FDACS would tell the public that 1900-ft was the recommendation of the USDA/ARS.

- Political support reasons. FDACS knew it had to pass certain legislation to empower the Department to conduct tree cutting in the backyards of residents. So, the Task Force was needed for political reasons, to show solidarity of both the citrus industry and regulators in the 1900-ft rule.

In June 2000, the International Citrus Canker Research Workshop was held, to prioritize future research for the task force. This certainly not part of Commissioner Crawford's directive for the group. The invited participants included scientists from the US and abroad, and members of the public. They were not permitted to ask questions about the ongoing eradication program in Florida.

Further information on the Task Force is provided on the online supporting documents website, including minutes of various meetings.

Organization of Task Force and Working Groups

The Task Force consisted of a three member executive committee and eleven supporting members. The Executive Committee members were Mike Shannon of the USDA/APHIS, Deputy Commissioner Craig Meyer of the Florida Department of Agriculture and Andy LaVigne, President of Florida Citrus Mutual (FCM). The Task Force did not include any residents. Nor were the lime grove owners of southeast Florida included, with the lone exception of Mike Hunt of Brooks Tropical. Brooks Tropical eventually joined the Broward lawsuit aimed at a court injunction against the program. A list of the initial members is provided in the online supporting documents website.

The Department did not freely distribute lists of members of the various working groups, dates of the meetings, minutes, and organizational structure, thus information on the working groups is sparse. The membership was also subject to change, so this information is only current as of years 1999 and early 2000. Only two academic researchers outside the Department participated on the Task Force, Dr. Browning and Dr. Graham, both from UF/IFAS. Dr. Browning's alternative was Dr. Peter Timmer. Drs. Browning and Timmer did not participate in the research efforts related to the 1900-ft field study and did not comment on the overall effectiveness of 1900-ft policy according to minutes of the meetings.

To further the Task Force objectives, subcommittees or working groups were formed in specific areas. The general responsibilities of these groups were determined through a review of the Task Force minutes. The June 30, 1999 Joint Meeting of the Science and Regulatory Issues Group was important because the 1900-ft policy was discussed as per the minutes of the meeting.

The citrus industry associations were well represented on the Task Force as the members representatives from Florida Citrus Mutual (FCM), Citrus Growers Association (CGA), Indian River Citrus League (IRCL) and the Highlands County Citrus Growers Association (HCCGA). George Hamner, representing the Indian River Citrus League was a strong supporter of the 1900-ft rule, provided there was a "risk assessment" provision, whereby grove owners could object to any eradication efforts.

- **Regulatory Issues Working Group** Lead by George Hamner, Jr., Indian River Citrus League (IRCL) and concerned movement of fruit out of quarantine areas. Some packinghouses were in quarantine areas, which caused regulatory complications. The citrus industry was concerned about USDA prohibiting sale of fresh citrus from quarantine areas to other states.

- **Science Issues Working Group (SIWG)** Lead by Mr. Tom Jerkins of Dole Citrus. Judging by comments at the Task Force meetings, it seems that the primary focus of this group was issues regarding sanitary practices and prevention of dissemination in the groves.

- **Public Relations/Education Issues Working Group** Lead by Lisa Beckman, FCM and Andy LaVigne, FCM, dedicated to informing others in the industry and public of the need to eradicate citrus canker.

- **Citizens Issues Working Group** This group was formed to help channel residents' concerns and was headed by Mr. Richard Gaskalla. This is the only group which had the participation of residents. The mayor of Hialeah, Julio Robiana, was a member of this group.

These working groups should not be confused with the Citrus Canker Risk Assessment Group, which was formed in 1995 to provide guidance to CCEP. They would periodically issue reports on their meetings. After their meeting on May 11, 1999, they issued their ninth report. This report is posted on the supporting documents website.

10. Excerpts from the 1999 Task Force Minutes

As discussed in the previous section, it is believed that the Task Force was not really to provide advice to the Department, as it had already decided on the 1900-ft policy, but was set up to sell it to the industry groups. The Department likely believed the public forum would be helpful in sidestepping rule making procedures. Excerpts from the meetings are provided below. Copies of the minutes and additional excerpts are provided in the online supporting documents website.

General Task Force Meeting, May 14, 1999

This meeting occurred just 3 days after the Citrus Canker Risk Assessment Group (CCRAG) met, and according to the minutes, voted to recommend the 1900-ft rule for residential areas within Miami-Dade and Broward counties. The group stated that eradication should restart using the 125-ft circles and as funds allowed, expand to 1900-ft circles.

Dr. Gottwald made a similar presentation on the field study at the Task Force meeting. The minutes, although very detailed on all other aspects of the meeting, provide no discussion of what was presented. A tabular listing was attached to the minutes, but only for Site 1.

Deputy Commissioner Craig Meyer raised the point that some evidence in Milton groves show that if eradication quickly follows discovery of an infected tree, the 125 ft rule is adequate (removes all secondary infected trees). Meyer stated:

> None of us can think that we can go to Dade County and cut 1900 feet immediately or in the near future. This would have the effect of pretty much removing all citrus from most of these neighborhoods... In addition to the impact from the citizenry down there, it we start to cut 1900 ft, we are not prepared to do that today and we don't have the resources.

Richard Gaskalla states that the Risk Assessment Group has an idea of applying the 1900-ft rule selectively to Broward County as it is the leading edge of the disease, and to create a two mile buffer zone in the south of the infestation in Dade County to protect the lime industry.

Ken Bailey, Program Director in Miami-Dade and Broward County, voiced strong concerns on the "public relations" issues, according to the meeting minutes:

> Ken said he has people that have guns pulled on them every single day. You can call it scientifically right if you want to, but from a public relations standpoint, you are not just going to walk on these properties and start cutting 1,900 feet.

Mr. Bailey perhaps was exaggerating a bit when he said "Guns pulled on them every single day" and likely just very concerned about the safety of the inspectors, particularly if the program dragged on. During the summer months, students from local colleges were recruited to make the inspections. The cutting would be done by contractors and the Department did not have a responsibility to their safety.

It is clear from this meeting, that Deputy Commissioner Meyer was reluctant to agree to 1900-ft without a more detailed plan which would include estimates of logistics and funding.

Science and Regulatory Issues Joint Meeting, June 30, 1999

At this meeting, a motion was made as follows: "Mandate that all citrus trees up to 1900-ft of a citrus canker positive tree be removed based on risk assessment." In the meeting's minutes, there is no mention of any discussion on the motion other than "Leon Hebb said he disagreed with this [1900 ft policy] because you would be taking out a lot of negative trees and this could be used against you in court." According to the minutes, the prior suggestions of a buffer zone did not come up at this meeting. Mike Shannon, USDA/Plant Health Board, commented:

> ... the program needs a specific recommendation not just the group saying remove 1900 feet when you need to, because that leaves the program open to legal action and makes defending their [the Department's] policies difficult.

General Task Force Meeting, July 16, 1999

This was a key meeting. Deputy Commission Meyer put forward the concept:

> ... the statutory law says that we have the authority to take out any citrus tree in Florida the Commissioner deems is at risk to harbor citrus canker. So we have the authority to go beyond 1900 feet, but beyond 125 feet. case law, we either have to have your cooperation or we have to pay you.

He was addressing the leaders of the citrus industry associations, not residents. Of course, no grove owner would be cooperative without compensation. Further, Leon Hebb puts the 1900 ft policy vote as a means of limiting the powers of the Department, stating, "If we wanted to go ten miles, we could do that. If you put the 1900 feet in there, you are just setting a limit."

Leon Hebb comment is surprising. Just two weeks earlier, he was worried about the legal ramifications if a large radius was used, causing a lot of "negative" trees to be removed. Now, at this meeting, he questioned why the task force should specify any radius.

Risk assessment was discussed as a procedure where the Department could exempt grove owners from the 1900 ft policy. The minutes of the meeting state:

> Richard Gaskalla answered (question was if an owner could appeal a risk assessment decision) "previously if the grower didn't agree with the risk assessment, then the technical advisory committee would address that."

The motion "Based on risk assessment, all citrus trees up to 1900 ft of a citrus canker positive tree may be removed" passed unanimously. This time Leon Hebb voted with the others. It is possible through risk assessment, grove owners could negotiate any eradication orders by way of a citrus industry friendly RA Group and Task Force Committee. Also, the wording is less direct, with the Department may remove trees and up to 1900 ft and not necessarily all trees within 1900 ft, leaving the option of a more limited circle. The vote by the 12 members of the advisory committee was only a recommendation to the Department and had no legal standing. Risk assessment is discussed in the next section.

General Task Force Meeting, November 16, 1999

According to minutes of the meeting, Dr. Gottwald presented a predictive model for the effect of storms, and concluded by stating, "Normal wind storm events can spread the disease 1900 ft." Richard Gaskalla commented, "if you go into a quarter section (160 acres), and you remove all the trees, that is a quarter section that you don't have to look back into." Craig Meyer also commented later in the meeting:

> Once we get the cutting done in that area, it is three quarters of a mile area we never have to look at again except periodic annual drive-by and pick up a citrus tree if replanted and the sprout issue we will need to deal with. By extending the eradication area, we will reduce the survey need.

The three quarters mile area is likely just an transcription mistake, and it is believed the Deputy Commissioner meant a third of a square mile or 213 acres, would be close to 260 acres enclosed by a circle with a radius of 1900 ft (1 acre = 43,560 ft^2).

There is no doubt that by November 1999, the task force was simply a means of convincing all participants, that the program would be successful in stopping the northern movement of citrus canker to the groves in Central Florida.

The next step was to change the laws of Florida, to empower the FDACS to implement the 1900-ft policy. Also, the legislature would have to authorize funding for the program.

General Task Force Meeting, February 3, 2000

It was clear in the November 16, 1999 meeting there was an anticipated cost effectiveness in cutting down trees rather than inspecting them. At the February 3, 2000 meeting, Mr. Gaskalla was asked how personnel requirements had changed from planned to actual. He responded:

> Richard Gaskalla answered, that actually, the number of people that we need particularly in South Florida has decreased and that is because we have changed our strategy away from intensified survey *[with a]* up to 1,900 feet destruction radius of citrus, so if you go in and find a tree and go out to 1,900 feet, there is no reason to go in there and loo anymore; you remove all that citrus, then that is an area we don't have to survey again. We are going to reprogram some of the survey money into control. As far as statewide resources, they are about the same level as they were.

Thus, the added cost of the 1900-ft program appears to be offset by reduced cost of inspections, indicating a clear cutting policy is cost effective.

Interestingly, the minutes of the Task Force meetings show nearly no discussion of the field study research, which supposedly was the foundation of the 1900-ft policy according to Department public relation officers. The minutes do not show any member asking when a report on Dr. Gottwald's research would be produced or if the study was reviewed by the Risk Assessment Group. When the interim report was sent from Dr. Gottwald to the Department on October 13, 1999, there was no mention of the report in any of the Task Force Meeting minutes provided by the Department (Oct 19, 1999, Nov 16, 1999, Feb 3, 2000, and April 11, 2000).

Also, the recommendation to use the 1900-ft rule in residential areas of Broward and Miami-Dade counties at the Risk Assessment Group meeting on May 11, 1999, but this recommendations does not appear in minutes of the Task Force until their October 19, 1999 meeting. Thus, Dr. Gottwald's claim that 1900-ft policy was decided not on the basis of any report, at least publically available, appears to be true.

11. Risk Assessment Policy

At July 16, 1999 Task Force meeting, when the 1900-ft policy was recommended, it was contingent on risk assessment evaluation for grove owners. A grove owner could seek a reduction in the cutting protocol based on this evaluation. It was also explained that this would not be an option for residents, because it was impractical. On June 30, 1999 at the Science and Regulatory Issues Meeting and on July 16, 1999 at the General Task Force Meeting, similar resolutions were passed related to the 1900-ft rule using risk assessment. The resolution passed by the Task Force stated that "Based on risk assessment, all citrus trees up to 1,900 feet of a citrus canker positive tree may be removed."

George Hamner, of Indian River Citrus League supported the recommendation as he states, "... you need to be able to go out this far because that is where we are showing potential movement averaged out per Dr. Gottwald's paper with tropical storms and weather patterns. We are trying to move the needle."

Risk assessment was a two level process, as shown below:

Level 1: Qualification checklist (6 yes answers qualify for a Level 2 risk assessment: (1) Is canker absent?, (2) Is disease incidence low? (3) Is the disease monocyclic? (4) Is the citrus isolated geographically from other citrus? (5) Are good sanitation practices being employed (6) Is survey unhindered? (7) Does property have adequate security?

Level 2: Factors considered: Residential or commercial, Cultivars/susceptibility, Tree size and age, Size of block, Tree spacing, Horticultural condition, Disease distribution in population, Weather events, Windbreaks present, Human activity, Lesion age and distribution, Disease severity, Tissues infected (leaves, fruit, twigs), Leaf miner activity, Program resources, Access/security, Compliance and timeliness, Management practices, Other citrus nearby, Other diseases nearby.

At the July 16, 1999 meeting, Mr. Richard Gaskalla felt including risk assessment to the 1900-ft rule would make the program more "bullet proof" (his words) if there were legal challenges according to minutes of the meeting (page 16). He was likely thinking about an individual grove owner contesting destruction of citrus, and the Department could respond that the action was based a thorough review of the removal action based on 7 qualifying criteria in Level 1 assessments, and 20 different factors, if Level 2 assessments were needed.

Ironically, risk assessment was more like the "last nail in the coffin" for the program. When the rules of the Department were challenged in Administrative Court, in July 31, 2001, Judge Laningham rejected the risk assessment, because it allowed exemptions to be made to grove owners based on subjective general factors, lacking any clear criteria. The factor which stands out the most, is "program resources." Presumably, if USDA ran out of compensation funds for grove owners, the resources would be too limited to cut the trees out to 1900-ft.

The Department had the option of revising drastically the risk assessment policy to conform to the requirements of Judge Laningham, with a set of clear criteria for each factor in risk assessment. Department rules can allow a more flexible policy, but they cannot exceed the mandates under Florida statutes. However, when the Department requested the legislature to embed the 1900-ft rule in a new Florida statutes, FS 581.184, to avoid the administrative court hearing, they lost all flexibility through Department rules to limit eradication efforts to anything less than 1900-ft.

12. Enactment of the 1900-ft Rule

Commissioner Crawford announced the 1900-ft rule on February 11, 2000 in an FDACS Press Release. A hot issue for grove owners was compensation for lost trees and the sale of citrus. This had been promised the grove owners using USDA funds, but approval or funding by Congress was incomplete. The tap had not yet been opened, but when the three programs were operational, the payouts were very generous to many. At the annual meeting of the Citrus Growers Association in December 2002, Dr. Jim Griffiths told the organization, "I suspect a lot of you would like to have canker" followed by, "Where is there a grove selling for $9,000/acre?" as reported by the Lakeland Ledger, December 5, 2002. The compensation was provided to keep the grove owners in business, replant citrus, but in many cases, citrus was never replanted.

The three programs for grove owners were: (1) Tree replacement re-imbursement, (2) Loss sale of citrus during the long period before new trees would be productive and (3) Crop insurance program (Florida Fruit Tree Pilot Crop Insurance Program). Commissioner Crawford stated that homeowners would receive gift certificates for their losses, which should not be mistaken for compensation. The problem apparently to offering compensation was that this was admission by the Department that something of value was being destroyed.

The Florida Statute 581.184 was passed in year 2000 adding to the authority of the Department to destroy infected and exposed trees. An exposed tree was defined as a citrus tree which harbored the bacteria but did not show symptoms of the disease. The 1900-ft distance was only mentioned in the preamble of the law. The law also contained a provision for the creation of a citrus free buffer zone. Since much of the concerned was the grapefruit and orange groves in the central part of Florida. The firewall concept at the time was to stop the northern advance of canker through a buffer zone, isolating the Miami-Dade outbreak from the groves in the northeast and central Florida. It had gained support from Dr. Griffiths at the November 1999 task force meeting.

The 2000 law stipulated that compensation had to be paid residents whose trees were within a citrus free buffer zone. No buffer zone was ever created. It is believed the Department did not go forward in creating a citrus-free buffer zone, because if these residents received payments, then those with exposed trees removed would likely also be seeking payments.

13. Key Documents and Articles

In this book, the key documents are, in general, public documents as provided by the FDACS and the USDA. Not all documents were readily available to the public. Documentation of the USDA payments to grove owners was provided based on Freedom of Information Act. Minutes of the Task Force Committee Meetings in year 1999 and 2000 were provided by the Department as required under Florida "Sunshine Laws."

Also, the Department was ordered to provide certain information in the Broward Court case in November 2000 in support of the 1900-ft policy. Prior to the court ordered release of information, the public knew very little about the research leading up to the 1900-ft program. The Citrus Canker Risk Assessment Group met regularly to jointly discuss and evaluate strategies to eradicate canker. The Department released a risk assessment report based on the May 11, 1999 meeting. Dr. Gottwald made a presentation on November 6, 2000 on the Florida field study. Copies of his view graphs are provided on the supporting documents website.

Two key documents were published on the Florida field study, a Letter to the Editor was published in Phytopathology in January 2001[14] and a more extensive article published in the same journal in April 2002.[16] In court documents, there are references to a "Gottwald Report" which may be referring to either of these two published articles or an internal report dated October 13, 1999.

Many documents on citrus canker are available from the FDACS website. From 2002 to 2004, three comprehensive articles were published as authored by Gottwald, Graham and Schubert, ,references 15, 20, and 26, respectively, which are listed at the end of this chapter for convenience. The Department in conjunction with USDA scientists held two research workshops in Florida. held, one in June 2000 and the second in November 2005 as follows: (1) International Citrus Canker Research Workshop, June 20 to 22, 2000, Ft. Pierce, FL, (2) Second International Citrus Canker and Huanglongbing Workshop, November 7 to 11, Orlando, FL. Links are provided to the abstracts and other information from these workshops.

Many excellent articles on citrus canker and epidemiology are available to the public from the American Phytopathological Society website. In researching the statistical tools used by epidemiologists, recent textbooks on plant disease epidemiology were consulted.

Finally, media stories from the Miami Herald, Sun Sentinel, Lakeland Ledger, Orlando Sentinel, Ocala Star Banner, and others were essential in providing details of the actions of the CCEP and a wide range of viewpoints on the program. Many of the older stories were obtained through Google's digital archive of newspaper articles as stored on microfilm.

Links to these workshop abstracts and transcripts are available from the FDACS website and the supporting documents website.

14. Concluding Remarks

"What choice do we have?" is likely how most Florida residents greeted an orange card hanging from their door informing them of a devastating disease call citrus canker was in their neighborhood. They were told that trees which had been "exposed" to citrus canker, would be destroyed. However, these exposed trees certainly looked healthy.

There was a standard Department's public relations pitch which went: "The 1900-ft is a science based policy. A USDA study showed bacteria land to the ground within 1900-ft of an infected tree, 95% of the time." Sometimes, they would add that this research in conjunction with years of field experience, resulting in the 1900-ft policy. Yet, in 1999, there was no research report, and the principal investigator, Dr. Gottwald, was not participating in any public relations effort. This was strange.

In early 1999, the Department sought the support of citrus industry for the 1900-ft eradication policy. The Citrus Canker Technical Advisory Task Force was composed of various industry and government groups. The Department seemed more oriented using the Task Force to gain support for the 1900-ft rule rather than having this group specifying an eradication radius. There would be just two options- support the 125-ft rule, which the Department was convinced would not work, or go with the 1900-ft rule.

The decision makers were not naive. Mr. Richard Gaskalla, Dr. Stephen Poe, Dr. Tim Schubert and Mr. Leon Hebb were veterans of Canker War II. They knew that some grove owners might end up suing the Department, particularly if there was insufficient compensation or a misdiagnosis of canker. It all had happened before.

Dr. Gottwald stated in a 2001 article that a consensus of attendees at a December 1998 meeting agreed on the 1900-ft policy. He testified to this in court, but the Department could not provide any evidence that the meeting took place or a list of the attendees to the meeting.

Further, in the minutes of the May 14, 1999 Task Force, there is practically no discussion of the field study. A table of results is attached to the minutes, but it is only for one site, and at the bottom of this table, it is stated the table was updated on May 26, 1999, so it is unknown what was presented at the meeting. Later, public relations officers would insist the 1900-ft distance was based on observations of how far citrus canker could travel, before it hit the ground.

Eradication policy has many aspects beyond the best technical policy. Based on minutes of the Task Force, members did not know of the legal consequences of the 1900-ft radius in terms of searches of residences without probable cause. Grove owners would not be concerned because no probable cause was needed for entry into commercial establishments, as the State had the right to regulate commerce. Deputy Commissioner Craig Meyer raised a concern that beyond a 125-ft, the case law allowing removals for exposed trees without compensation might no longer apply.

By now, this chapter has whetted the appetite of the reader to know more. In the next chapter, prior attempts at eradicating citrus canker are described. Most plant diseases are not eradicated, they are simply brought under control or made less damaging.

There is much more information on the online supporting documents website. The short notes further extend many of the topics in this chapter. Images of citrus canker and similar looking diseases are also posted in the website.

Short Notes on the Website

SN 1.1 Why this Website?
SN 1.2 Dr. Whiteside's Contribution to Citrus Canker Research
SN 1.3 Risk Assessment Group and Task Force
SN 1.4 Worldwide Distribution of Citrus Canker
SN 1.5 Okeechobee Saga
SN 1.6 What is an Exposed Tree?
SN 1.7 Eradication Delays and Related Statistics
SN 1.8 Overwhelming Emotions, by Patricia Haire
SN 1.9 Citrus Canker and the Giant Yellow Swallowtail, By Roger Hammer

Conversions and Abbreviations

1 acre = 43,560 square feet
1 section = 1 square mile = 640 acres

ACC = Asian citrus canker
DOACS = Florida Department of Agriculture and Consumer Services (old acronym)
FDACS = Florida Department of Agriculture and Consumer Services
FDACS/DPI = FDACS, Department of Plant Industry
CCRAG = Citrus Canker Risk Assessment Group
FCCTATF = Florida Citrus Canker Technical Advisory Task Force, referred to as the Task Force
ICCRW = International Citrus Canker Research Workshop
UF/IFAS = University of Florida, Institute of Food and Agricultural Sciences
USDA = United States Department of Agriculture
USDA/ARS = USDA, Agriculture Research Service
USDA/APHIS = USDA, Animal and Plant Health Inspection Service
USDA/APHIS/PPQ = USDA/APHIS, Plant Protection and Quarantine

Key Documents (all references are provided at the end of the book)

15. Gottwald, T.R, Graham, J.H., Schubert, T.S., 2002. Citrus Canker: The Pathogen and Its Impact, Plant Health Progress, published online at www.apsnet.org (official website of the American Phytopathological Society) .

20. Graham, J.H., Gottwald, T.R., Cubero, J. and Achor, D. S., 2004. Xanthomonas axonopodis pv. citri: factors affecting successful eradication of citrus canker, Molecular Plant Pathology, 5(1), 1-15.

25. Schubert, T.S. , Gottwald, T.R., Rizvi, S.A., Graham, J.H., Sun, X., Dixon, W.N., 2001. Meeting the Challenge of Eradicating Citrus Canker in Florida- Again, Plant Disease, Vol. 85-4.

2. Four Decades of Eradication Programs

> 1915, Stevens projected, without providing any convincing data, that if canker persisted it would be uncontrollable and even kill trees.
>
> The measures that are currently aimed at canker eradication are costly, laborious and the prospects for success are extremely limited.

Dr. J. O . Whiteside, former Research Plant Pathologist at the University of Florida, IFAS, Citrus Research Education Center on Canker War II, as published in April 1988.[17]

> At stake is The Supply of Fruit to Half the World

Palm Beach Post, 1915, Editorial Headline

> The past is never dead. It's not even past.

William Faulkner, Requiem for a Nun, 1951.

---

1. Eradication Attempts by Florida Department of Agriculture

Before the 1900-ft rule was implemented, regulators were likely thinking, how important it was not to repeat mistakes of the past. By year 2000, citrus canker had proven to be an elusive and persistent disease, surviving in Florida for approximately 90 years. In two major eradication campaigns against citrus canker, the Department had tried to eliminate the disease by burning down both infected and healthy trees primarily in nurseries, and to a lesser degree in groves. In Canker War II, there was destruction of both residential trees, but on a very limited scale.

The three major campaigns are divided into five eradication programs as summarized below. The last eradication campaign, Canker War III, is divided into two programs, because the policy changed in year 2000 from the 125-ft to the 1900-ft removal radius. Also, Canker War II is divided into two programs — false canker and true canker eradication programs, which overlap in time.

1) Canker War I (1912 to 1933) Canker, discovered in 1912, was thought to have come from infected nursery stock from Japan. Eradication program began in 1915. Citrus groves in Miami-Dade County were burned. Approximately 3.3 million trees were destroyed. Citrus canker declared eradicated in 1933.[12]

2) False Canker War (1984 to 1989) The Department destroyed millions of nursery plants to control a foliar disease, which they had incorrectly identified as Asian citrus canker. The program devastated the nursery industry as inspectors tracked and destroyed potentially infected plants from one nursery to the next.

3) Canker War II (1986 to 1992) While attempting to eradicate false canker, inspectors discovered true citrus canker. The six year campaign was a much smaller effort than Canker War III. Citrus canker was declared eradicated in year 1994.

4) Canker War III Part 1 (1995 to 2000) Consisted of a hatracking policy, then 125-ft cutting policy, followed by a cutting moratorium, then a resumption of 125-ft policy.

5) Canker War III Part 2 (2000 to 2006) The 1900-ft radius cutting policy was implemented for groves, nurseries and residential areas. This final attempt to eradicate canker, began in year 2000, with the 1900-ft rule, is reviewed in Chapter 1.

In total, in the 94 years between 1912 to 2006, four decades were spent attempting to eradicate citrus canker. Citrus canker is still very much present in both the groves and residential neighborhoods today. The online supporting website provides more historical documentation.

2. Canker War I (1912 to 1933)

This brief account of Canker War I relied on articles by Schubert et al[12], Whiteside[15,16,17] and Stevens[13] as referenced at the end of this book. Also, newspaper articles retrieved from the microfilm from the Broward Historical Commission by Sue Peterson aided in this account of events.

"At Stake is The Supply of Fruit to Half the World" read the headline in an editorial in the Palm Beach Post in 1915. Half the world's supply of fruit was not going to disappear from the world. However, a war on canker had been declared, and as the saying goes, the first casualty of war is truth. The devastating disease or killer canker labels were started in the early 1900's, based on exaggerated claims of the harm citrus canker cause to citrus trees.[16,17]

The first canker war began by eliminating citrus canker primarily from nurseries around Miami-Dade County by burning them down.[12] Kerosene flame throwers from fuel trucks were the weapon of choice. Canker was found in groves on both east and west coasts of Florida. Dr. Timmer, former Professor of Plant Pathology at UF/IFAS notes:

> They took really drastic action. They burned everything, nurseries, and groves. And they basically had people running around with flame throwers, dousing the trees with gasoline and burning them on the spot. And they disinfected everybody with mercuric chloride. I'm amazed some people lived through that.[7]

Ultimately, citrus canker was discovered in 26 counties in Florida, so the epidemic was about as wide spread as the most recent outbreak (Canker War III). Approximately 94% of all trees cut (3 million trees) were from the nursery and the remainder 6% (257,745 trees) were from the groves.[12] No residential trees were destroyed.

Once canker was discovered in a nursery, part or all of the nursery was destroyed by burning. In groves, it was mandatory to destroy all citrus trees with visible lesions.[17] Some grove owners elected to remove healthy trees in the immediate vicinity of an infected citrus tree.[16] Maps showing the distribution of canker are provided in the website.

While the eradication program went on for 21 years, it appears that most of the discoveries were made in the groves in the first five years (1912 to 1916).[15] Whiteside[16] attributes this early success to the thorough inspections and massive eradications in nurseries. It is likely nursery stock was very short supply by the end of 1916 with 3 million trees destroyed.

Records from the newly created State Plant Board show that before restrictions were imposed in year 1915, 338,512 trees were moved from infected nurseries and planted in groves in 21 different counties.[17] Regulators tracked down, and destroyed by burning approximately half of these trees. State records do not show what became of the other trees from infected nurseries.[17]

After the massive destruction of planting stock, new discoveries were made in the groves on occasion from 1920 to 1927. It is possible, that the introduction of infected plants into the groves occurred in for several years prior to their discovery, but the lesions were too small or not prevalent enough on the trees to be noticed.[17] Canker often begins with a few pinhead sized lesions in new flushes are at the top of the canopy and are often missed by inspectors.[17]

Citrus canker was traced back to a contaminated shipment of seedlings from Japan around 1910, with trifoliate rootstock.[1] It is possible that these imported citrus seedlings were in needed to replenish nurseries following the devastating hurricane which passed through the agricultural areas of Miami on October 18, 1906.[17] Citrus canker was present in six states in addition to Florida— Texas, Louisiana, Alabama, Mississippi, Georgia and South Carolina. Eradication of canker occurred in all seven states and the US was considered canker-free by the USDA in 1947.

Early in the eradication program, exaggerated claims of the harmful effects came from state officials. It is likely these claims were made to gain financial support for the eradication program.[17] According to an article by Dr. Whiteside, "In 1915, Stevens projected, without providing any convincing data, that if canker persisted it would be uncontrollable and even kill trees."[17]

Funding of the canker program was controversial in the legislature.[9,10] Representative Mathis, as reported in the Palm Beach Post in May 25, 1919 stated, "In 1915, they came here asking for $150,000", he said, "and now they declared that, if we would give them that, they would not ask for more money. The next session, they are back asking for $300,000 — now they are back for an additional $175,000." To pass the 1919 canker appropriation bill, it became necessary to gain favor with representatives from other counties. An additional $10,000 was included to protect the honey bee industry and another $25,000 was tagged on to the bill for tobacco diseases. The debate in May 1919 over canker funding became extremely divisive and almost broke Florida into two states.[10]

In 1933, there was an official declaration that Florida was canker- free. As commented on by Dr. Whiteside:

> Surprisingly, there are no records of anyone ever challenging the conclusion reached in 1933 that all the canker outbreaks in Florida had been found and that the disease was, therefore, eradicated. Yet, if one considers the nature of the disease, its epidemiology, and the difficulties of thoroughly inspecting the foliage for canker lesions, particularly on larger trees, it seems unreasonable to believe that canker disappeared from Florida because of the eradication efforts alone. Did canker really disappear in Florida by 1933 or did some remnants of the disease remain after this time?[17]

Dr. Whiteside was skeptical of the claim of complete eradication within the groves:

> There must have been many instances where canker pustules were missed by the inspectors. The human eye would never be able to detect the last vestiges of canker infection. Canker pustules, particularly the smaller ones are not very conspicuous and it would take one surviving pustule to perpetuate the disease. Pocket of citrus canker may still have existed in some groves even after the pathogen was supposedly eradicated. Perhaps the disease has survived in some locations even to this day *[1986]*. Small amounts of leaf or fruit spotting caused by canker could be missed by even the most ardent observer.[17]

Note, pustules are essentially the same as lesions, with respect to citrus canker.

The complete lack of discoveries in the time period from 1929 to 1933, may be related to a change in priorities due to the Great Depression. Did the State of Florida really want to be discovering "killer canker" in its groves and nurseries, at a time where there were severe trade embargos from our European trading partners? Foreign trade from 1929 to 1933 dropped 70% (see wikipedia.org/wiki/greatdepression)

3. False Canker War (1984 to 1989)

Nearly five years were spent attempting to eradicate a disease that looked similar to citrus canker. It was initially called "nursery strain" canker, because it was discovered in a nursery in Polk County. The lesions looked similar to true citrus canker (ACC), particularly in its early stages, however the lesions in nursery canker were flat, not raised.[12] The lesions form on the leaves and twigs. As scientists later learned, lesions from nursery strain are rarely seen on the fruit, and has only been observed on citrus grown on flying dragon trifoliate rootstock.

The first discovery of nursery strain canker occurred on August 27, 1984 at Franklyn Ward's nursery in Avon Park, Highland County. The owner was Franklyn Ward, and his father had started the nursery in 1928. All citrus in Florida is grown as grafted plants, and Ward's nursery sold other nurseries budwood to make the grafts.[5] Ward had kept meticulous records of all sales, and helped track down every nursery who bought his plants. George Ward quickly rebuilt his nursery and obtained registration for his budwood.[5]

The Department's canker eradication rules called for the destruction of any nursery, which had received plant materials from another nursery where citrus canker had been found. Ultimately, this lead to the destruction of 55 nurseries. Ward's 60 acre nursery contained approximately 1 million citrus was one of the first nurseries to be destroyed. The William's nursery, with 45 acres of citrus, was discovered on September 18, 1984 with nursery strain citrus canker and subsequently destroyed. The Department burned down A. Duda and Sons' 50-acre nursery with 500,000 seedlings, because the nursery had received stock from Ward's and William's nurseries. One after another, the nurseries on the west coast of Florida were burned to the ground.[8]

By September 21, 1984, there were 50 nurseries under quarantine.[8] A quarantine for a nursery essentially puts the nursery out of business because the plants can not be sold. A total of 17 commercial nurseries went out of business in the first 16 months of the program. It is likely these drastic measures were taken as regulators feared wide spread dissemination of citrus canker throughout the groves, leading to interstate quarantines and trade embargos of all fresh fruit.

The eradication program just seemed to go from bad to terrible in the next few months. On November 1, 1984, the William's nursery had sold 480 citrus seedlings to Colombia.[4] Additional plants may have been sent to Costa Rica.[4] The USDA advised these countries to destroy their plants. The situation was made all the worse, by the constant references in the news media to the highly contagious "killer-tree" canker. In addition, the Department was also asking residents to voluntarily destroy 89,000 citrus trees purchased at commercial outlets, which had bought their citrus from the William's nursery.

Many nursery owners were furious at the Department, for what they considered to be an over reaction. Grove owners wanted restrictions lifted against shipment of citrus from their groves. By January 1, 1986, the Commissioner of Agriculture Conner Doyle took measures to partially lift the quarantines. In March 1986, Dr. Robert Stall, the chairman of the joint state and federal technical advisory committee resigned stating in the media that the job had been very stressful, balancing the conflicting interests of citrus industry representatives, government officials and plant pathologists.

The USDA/APHIS wanted to know where this new strain of canker originated, likely with the intent of declaring embargos against the country of origin. In October 1984, Dr. Civerolo indicated the strain was not matching others from Mexico and other countries with canker. Fresh citrus coming from Mexico had to be dipped in a chlorine solution to kill any bacteria. An embargo had existed on citrus fruit from Mexico when citrus canker had been discovered there in July 1982. It was lifted to non-citrus producing states in January 1983 and a completely lifted in November 1983. It is likely some of the grove owners wanted to blame Mexico for the outbreak and re-impose the trade embargo on Mexico.

As scientists began a more in-depth examination of canker symptoms, questions were raised to the proper classification of nursery strain canker. Microbiologists could see many similarities between the nursery strain and the more familiar Asian citrus canker. Dr. Edwin Civerolo, a plant pathologist at the headquarters of the USDA/ARS in Beltsville, MD was sent samples of the nursery strain. In December 1984, he stated that he was 99% sure the disease was citrus canker, but unsure of which particular strain of canker. It was called citrus canker strain "E" at the time.

In July 1986, Dr. Sterling Long, a microbiologist with a private laboratory in Cincinnati, Ohio, expressed serious doubts about the Department's claim that nursery citrus canker was simply a different strain or form of Asian citrus canker. The disease did not seem as contagious as Asian citrus canker as lesions were not seen on fruit rinds in the groves. Dr. Long had a long career as a researcher at the Citrus Research Education Center in Lake Alfred, Florida. The Department refused to provided him any samples. However, he finally convinced the USDA to provide him with a sample of the bacterium and confirmed the nursery canker was not another form of Asian citrus canker, but a distinctly different pathovar.

If the pathogen were biologically different from Asian citrus canker, then was it as harmful? This question would come under court scrutiny, as scientists testified in December 1987 in the Polk nursery lawsuit. Both Dr. Civerolo and Dr. Brlansky of UF/IFAS testified in court that the disease appeared to be a less aggressive form of canker, as lesions that did not expand in size like Asian citrus canker (referred to as the "A" strain of canker).

With more research, other scientists came to similar conclusions as Dr. Long. Drs. Dean Gabriel, Professor of Plant Pathology, University of Florida, and John Hartung of the USDA/ARS in Beltsville, MD, concluded that the nursery canker was uniquely different pathovar, requiring a different scientific nomenclature. In early January 1988, Dr. Gottwald concluded nursery canker would not threaten the Florida citrus industry nor spread to other states. He urged APHIS to allow interstate shipment of citrus fruit. He blamed the continued quarantines on the USDA/APHIS acquiescing to the other states for protection of their fruit.[11] Dr. Gottwald worked for the USDA, however his branch of the USDA, the Agricultural Research Services (ARS), did not set quarantines or eradication policies. By January 1988, even Dr. Robert Stall, the former head of the Citrus Canker Technical Advisory Committee, was agreeing that the nursery strain was a less aggressive disease and the quarantines should be lifted.[11]

For nearly four years from 1985 to 1989, nursery owners armed with the opinions of experts clashed with regulators from the Florida Department of Agriculture as to whether there was cause to destroy their nurseries due to "nursery strain" canker. These clashes were quite public, occurring at Task Force meetings and in the courts. It would not be in the interest of officials with FDACS to admit that for years, what they were destroying was not citrus canker.

In February 1989, Commissioner Conner Doyle ended the eradication of nursery strain canker, but his popularity among nursery owners and some citrus industry associations was poor. Dr. Whiteside of the UF/IFAS published three articles, chronicling the history of exaggerated claims of the destructiveness of citrus canker disease and the difficulties in proper identification.

Complete de-regulation of nursery strain canker occurred in 1990. In the 1990 elections, Conner Doyle, a Republican, lost to Senator Bob Crawford, a Democrat, for a four year term.

Today, the nursery strain canker is referred to as citrus bacterial spot (CBS) disease, with the scientific name *Xanthomonas axonopodis* pv. *citrumelo*.[11] No controversy exists today as to whether the nursery disease is distinctly different pathovar, *citrumelo* instead of *citri*. However, controversy still exists as the whether inspectors can accurately determine the difference between CBS and citrus canker strictly based on visual inspections in the field, particularly at early stages and in less than optimal environments such as residential areas. CBS is rarely found in groves[3] but its prevalence in residential areas is unknown.

Additional articles on CBS including microbiological research are posted on the website. It has shown that the flat lesions is due to a lack of hyperplasia (rapid cell division within the mesophyll layer causing a crater like lesion at the surface).[3] Further, CBS has only been located within the state of Florida as of year 2011, so the origins outside of Florida are unknown.[3]

The conflicts between the Department and scientists during the nursery canker saga and subsequent law suit were documented in an article entitled "The Big Lie" by Philip Longman.[6] His article alleges that when the Department was challenged in court, officials continue to insist the nursery strain could be considered another form of Asian citrus canker, requiring eradication, even though they knew better.[6]

4. True Canker War II (1986 to 1992)

Concurrent with the eradication of false or nursery canker, true canker or Asian citrus canker was discovered in residential area of Pinellas County in 1986 and later, in June 1986, on Anna Maria Island, Manatee County. The Director of FDACS/DPI, Sal Alfieri stated that he feared lawsuits from residents that could, in his words, "bring the program to a screeching halt", but other officials stated that residents were cooperative. A zone of approximately 2,000 citrus trees showed 232 canker infected trees. An 27 acre area of the Manatee Fruit Co. Grove, about 7 miles northeast of Anna Maria Island was infested with citrus canker. This time Director Sal Alfieri had definitive lab tests confirming the disease was Asian citrus canker. The means by which the canker was introduced to the groves and residential areas was never established, but one possibility was the plants came from a nursery which had been destroyed during the prior two years of battling nursery strain canker.

Since eradication had begun in 1912, the standard procedure was to burn both healthy and infected trees. This had to be changed in 1986, as residents would not exactly appreciate their backyard trees being doused in kerosene and then lit on fire. The trees were cut down and hauled away to be burned. Assistant Director of FDACS/DPI, Richard Gaskalla, stated the cooperation from residents was excellent. In the groves, the Department destroyed all infected trees by burning and defoliated all other trees within 50-ft of the infected tree. This was a less aggressive policy than the 125-ft rule as implemented in beginning of Canker War III.

The eradication effort lasted from 1986 to 1992 and overlapped two years with the False Canker War. A declaration of eradication was made in 1994 following two years of no new incidences of citrus canker in the Manatee area. A total of 87,741 commercial citrus trees on 594 acres and 600 residential trees were destroyed.[16] The concern of lawsuits from residents never materialized, likely because so few backyard trees were eradicated.

The cost of Canker War II, including both the false canker war and the true canker war, as reported in the Miami Herald on January 21, 1994 was 200 million dollars. It was also reported that there was an additional 100 million dollars paid to the nursery owners who lost their trees, so the full cost may be 300 million dollars.

Citrus canker discoveries were limited to the west coast of Florida. There is no indication that any other areas were subjected to inspections. Citrus canker was discovered in residential citrus in Hillsborough, Pinellas, Sarasota and Manatee counties.[1] Discoveries also include two Manatee County commercial groves and one commercial grove in Highlands County.

Was the outbreak of citrus canker in 1986, caused by reintroductions from contaminated nursery trees or holdovers from the prior canker war? Perhaps a combination of both were responsible for the outbreak. If there were holdovers or pockets of infection trees where the canker disease would find hosts and spread, it would likely be in small, poorly managed or abandoned groves or nurseries which had not been inspected. A small backyard and/or unregistered nursery would also be an ideal environment to disseminate citrus canker for decades and spawn more pockets of infected trees. Trees with canker symptoms would simply be pruned back before they were sold.

Although Florida was officially declared canker-free on January 20, 1994, many questioned whether canker was truly gone. Evidence suggests it was a case of curtailing inspections, based on lack of funds, as stated in an USDA-APHIS/PPQ 1999 Environmental Assessment Report:

> A localized outbreak of bacterial citrus canker in Manatee County *[part of Canker War II]* was declared eradicated, but no funding was provided for further survey of the area despite the recommendation from USDA and FDACS to continue surveillance of the high risk areas. Bacterial citrus canker was again detected in Manatee County in 1997 and the genetic profile of the detected bacterium indicated that was not eliminated in the earlier infestation.[14]

Dr. Stephen Poe of the USDA/APHIS likely prepared this report as he is listed as the contact agent. He was a key participant in both Canker War II and III. In Canker Wars III in his position as Operations Officer, Program Support Staff, USDA/APHIS/PPQ, he reviewed claims for reimbursement submitted to the grove owners and nurseries.

5. Canker War III, Part 1 (Oct 1995 to Jan 2000)

Canker War III officially began in October 1995, just 21 months after being declared officially eradicated. Citrus canker was discovered in September 25, 1995 in residential area of Miami-Dade County (Sweetwater) approximately 5 miles from Miami International Airport. The discovery came as a result of a backyard inspection by a Medfly inspector.[12]

Hurricane Andrew in 1992 may have contributed to the epidemic. As stated in the article by Schubert et al:

> Hurricane Andrew hit South Florida in 1992 about the time that the disease made its appearance, and some have speculated about the storm's possible role in the introduction. Several other exotic pests appeared in South Florida at about the same time: the Asian citrus leaf miner, the brown citrus aphid and the Asian citrus psyllid."[12]

The Asian citrus psyllid is a vector for citrus greening and the brown citrus aphid is the vector for citrus tristeza virus. By April 17, 1997, the brown citrus aphid had been discovered in 20 counties in both the east and west coast of Florida according to a FDACS press release.

Canker War III, from October 1995 to January 2000, there was limited cutting to a radius of 125-ft from an infected tree. In Table 2.1, the eradication protocols used during this period, and the month of cutting are shown. In total, cutting of citrus trees lasted for a total of only 19 months.

Table 2.1: Eradication Protocols from Oct 1995 to January 2000

Dates	Protocols Employed	Months of Cutting
10/95 to 1/97	Hatracking infected tree	0
1/97 to 2/98	Initial 125-ft Policy	13 months
2/98 to 6/99	Moratorium on Healthy Tree Cutting	0
7/99 to 1/00	Resumption of the 125-ft Policy	6 months

Hatracking protocol (Oct 1995 to January 1997) and initial 125-ft policy

From the beginning of eradication in October 1995 until January 1997, all citrus tees within 125 ft of an infected tree were pruned back to brown wood (hatracking)[12] and the infected trees were removed.

In January 1997, a cutting program was initiated for 13 month lasting until February 1998. A moratorium period was declared by the Commissioner in February 1998 and lasted 13 months. No healthy trees were cut in Miami-Dade County during this period. From July 1999 to January 2000, a 125-ft policy period was resumed. Then on January 1, 2000, the cutting radius was changed to 1900-ft. This policy had a major impact on the cost and level of destruction of healthy (symptomless) trees.

The hatracking process seem good in principle, but difficult in practice. It requires that all nearby infected trees to be hatracked at the same time, so the area would be free of disease sources. However, due to latency (time to develop symptoms) of the bacteria, it is difficult to be sure an area was free of disease. Re-inspections would be necessary. Also, as the trees grew back, the new flushes on the trees would be very susceptible to canker.[12]

In January 1997, the policy changed to removing both the infected tree and healthy trees. A herbicide was used in an attempt to prevent healthy trees from growing back. By February 1997, there were 88 positive residential sections (section = 1 square mile or 640 acres) in Miami-Dade County. In the period from January to May 1997, the Department eradicated 2,400 trees in 20 weeks, which is an average of 120 trees per week.

The slow pace of eradication may be related to under funding of the program. The greatest deficiency in this period was the lack of inspectors and cutting crews. The total cost of the program from October 1995 to January 1997 was about 9 million dollars.[12]

In 1997, citrus canker was discovered in Broward County and in the lime groves in Homestead. However, in an article published in August 1997, researchers from FDACS and the USDA, indicate that the incidences of citrus canker were decreasing. But, the cases of citrus canker were scattered over a larger areas. This made inspections more difficult.

Also, in 1997, citrus canker was also discovered on the west coast of Florida in Manatee County. There seemed to be general optimism that this was a limited discovery, and perhaps contained with the 125 ft rule. A status report from the NAPIS pest tracking website, for the week of October 13, 1997 states , "Within two weeks, eradication activities will be complete in Manatee County. " Eradication of all known infected plants and those in close vicinity of these plants certainly does not mean containment of the disease.

February 26. 1998: Commissioner Crawford announces changes

The Commission of Agriculture Bob Crawford announced on February 26, 1998 a series of changes to the CCEP. The changes consisted of a one year moratorium on exposed tree cutting. The reason for the moratorium as stated in his press release, was the acceleration of infected tree removals.

Other publications claim that the Commissioner enacted the moratorium due to negative reactions of residents.[2, 12] Dr. Graham, professor of plant pathology at University of Florida, wrote in 1998:

> The eradication agencies are even more hampered than ever by groups of residential property owners who have legally impeded access to their property for survey and successfully lobbied for a moratorium on destruction of exposed trees.[2]

Dr. Graham's statement misconstrues the actual situation. The entry onto property was lawful in 1998, since no court had ruled otherwise. If a resident attempted to block entry, the Department could call on local law enforcement, and the resident would be subject to arrest. More likely, in 1997 to early 1998, both USDA or FDACS officials were growing concerned about funding a program which could result in lawsuits, charging illegal entry and inadequate compensation. Canker eradication programs and contentious lawsuits just seem to go hand-in-hand.

The Valera case (*Dep't of Agriculture v. Valera*) began in November 1997 when attorney John Ruiz representing private residents, filed a lawsuit claiming that compensation was required for destroyed healthy trees. The class action lawsuit was not seeking to shut down the program, but to obtain compensation for residents with eradicated healthy trees. The lawsuit was certified as a class action lawsuit by 1998. The Third District Appellate Court in May 1999 dismissed the case, citing the exposed trees had no value. With the dismissal of the Valera lawsuit, the Department had avoided the first legal challenge to the eradication program.

Commissioner Bob Crawford's program changes in February 1998, called for a one year experiment to track the spread of canker from infected tree to expose ones. The "experiment" is referred to in this book as the "Florida field study." Additionally, Commissioner Crawford proposed a program to replace the trees which had been destroyed, which was termed a "reforestation program." This program which would provide residents of with citrus replacement tree. The re-forestation program would begin after a two year period of no new discoveries. The reforestation plan was quietly ended and no citrus trees were ever bought by the Department.

Restart of the 125-ft Cutting Program, June 1999

In April 1999, the Varela lawsuit was dismissed. In May 1999, there was an unsuccessful attempt to have the case heard by the Florida Supreme Court. In June 1999, Commissioner Crawford announced a restart of the 125-ft cutting program to include both healthy and infected trees. This lasted until January 2000 when the 1900-ft policy was implemented.

The restart of the 125-ft program may seem odd, given the Citrus Canker Task Force seemed to be close to recommending the 1900-ft protocol. However, the Department received in April 1999, 40 million dollars from federal sources, as part of a program to control invasive plants and animals. Thus, it was likely this would continue to fund the program, until the Florida legislature could provide additional funds. So, in 1999, a 125-ft protocol was within their budget, but a 1900-ft protocol was likely too expensive.

Prior to the resumption of cutting, there were two town hall meetings as a result of the newly formed Citizens Issues Working Group. The first meeting occurring on June 18, 1999 was held in Miami Springs, Florida which is a small community north of Miami International Airport. The press and public were invited. The Miami-Herald reported only about 10 people showed up for the first meeting. According to the Miami-Herald story on June 19, 1999, Dr. Tim Schubert, Administrator of the Plant Pathology Department in FDACS/DPI stated that recent research concluded that a radius of 1200-ft would eliminate 99% of the disease, while the 125-ft radius would eliminate only 30% of the disease.

On January 1, 2000, Commission Crawford enacted by Executive Order, the 1900-ft policy, which began Canker War III, Part 2. This greatly expanded war against canker has been summarized in Chapter 1.

6. Concluding Remarks

Florida's Commissioner of Agriculture and Consumer Services Bob Crawford in 1995, had a critical decision to make. Should the Department once again declare war on citrus canker? Or was a more peaceful canker management program a better option? Only Canker War II had cut down residential trees, and this was on a very limited scale. Both canker wars resulted in massive destruction of nursery stock. Both wars were highly contentious and ultimately unsuccessful.

Canker War I seemed successful, once completed. However, kerosene filled trucks with flame throwers charging through nurseries, was not going to be part of the Commissioner's game plan! One ought to be skeptical about the 53 year canker-free (1933 to 1986) era. Why would any grower want to let the Department of Agriculture know that they had canker? There was no way they could be compensated for losses. And they would risk trade embargos with other countries.

Canker War II began with the destruction of a disease that was incorrectly diagnosed as a new strain of citrus canker. However, midway through the program, real canker was discovered. Canker War II came to an end, as Dr. Poe noted, because the Department decided to limit inspections in high risk area. At least, Canker War II could end on a positive note, with declaration of a canker-free Florida.

The Commissioner most likely wanted to avoid the contentious disputes during Canker War II. The most knowledgeable research plant pathologists from the University of Florida and the USDA/ARS contradicted the Department's claim that "nursery canker" was just a different form of true citrus canker. Further disputes between the Department and the USDA occurred in 1988 when the USDA refused to lift the quarantines for interstate shipments from canker infected groves. There were bitter disputes both in court and during task force meetings.

Perhaps the lesson left behind by Canker Wars I and II was that it was possible for the Department to become unpopular among the citrus industry leaders and particularly the nursery owners, if they went too far. Going forward, it is believed the Department understood that a generous compensation from the USDA was necessary component of an eradication program.

Short Note on the Website

SN 2.1 Additional notes on Canker War I

3. CITRUS CANKER BIOLOGY

> Citrus canker is highly contagious and can spread rapidly by windblown rain, people carrying the infections on their hands, clothing or equipment and by moving infected or exposed plants or plant parts.

Florida Department of Agriculture/Division of Plant Industry website, November 2000

---

1. The Elusive Bacterial Disease known as Citrus Canker

For more than 90 years, the infection mechanisms and life cycle processes have been studied by scientists around the world. Knowledge of the life cycle requirements aids in identifying the effective measures to control or eradicate the disease.

Citrus canker creates lesions on the leaves, stems and fruit of citrus trees. The bacteria are released from these lesions. The infection process requires wet conditions, from either rain or an irrigation system, for release of bacteria from a lesion and later for the entry into the pores of a leaf, stem and fruit of a citrus tree. Only plants of the Rutaceae family act as hosts for the disease.

Many factors cause a citrus tree to be more susceptible to canker infection and promote disease dissemination. These factors include the resistance of the species/cultivar, ambient temperature, weather conditions and planting density. For example, grapefruit are generally more susceptible than kumquat trees. Gravitational force on rain droplets limits travel distances. A light rainstorm is more likely to transport bacteria further more than a heavy rainstorm.[27]

The dissemination of canker requires a means of transport. Wind is one form of transport demonstrated by scientific experimentation. Unfortunately, there are numerous mechanisms of transport suggested in various publications that are unsupported by experimental studies. For example, it has been stated that birds can potentially spread the disease, without a single experimental study. Others suggest that insects may be vectors or carriers of citrus canker. These unsupported mechanisms are reviewed in this chapter.

During both Citrus Canker War II and III, opponents of the programs questioned if citrus canker was as harmful as claimed by FDACS. Media reports during Canker War II, referred to "killer canker." In the final outbreak, media reports referred to the disease as devastating to citrus trees. One dictionary defines "devastate" as, " to destroy much or most of (something) : to cause great damage or harm." The Department has not shown evidence of trees dying of citrus canker anywhere in Florida.

For citrus canker and other diseases, there have been quarantines placed on other countries, particularly Mexico and Argentina, prohibiting the import of fresh citrus fruit, because the citrus has been grown in an area where citrus canker is present, even though the fruit has been sanitized and inspected.

A 2007 USDA/APHIS study concluded the movement of inspected and processed citrus has a very low risk of disseminating canker.[37] A draft report, prepared as for the Citrus Canker Risk Assessment Group, and posted on the FDACS website had come to very similar conclusions, on July 14, 1999. [42]

This summary provides basic biological information for understanding the Florida epidemic. It is not meant to be a comprehensive review of all aspects of citrus canker. In recent years, microbiologists have investigated the infection process at a cellular level, describing the infection process in terms of microbe-host interactions. This research is highly technical and beyond the scope of this book. However, this research is important as it may lead to development of disease resistant cultivars. Links to publically available articles are provided in the online supporting documents website.

2. Infection Mechanisms

Bacteria enter the plant tissues through the stomata cell opening, which are located on the underside of the leaves and on twigs and fruit. The stomatal number in oranges is estimated to be approximately 190,000 to 240,000.[2] The numerous stomata cell opening might suggest easy pathways for which bacteria can enter, but some studies have suggested that the guard cells controlling the opening can actually sense the presence of pathogens. Entry can also occur through wounds in plant caused by insects or blowing sand or other particles.[25]

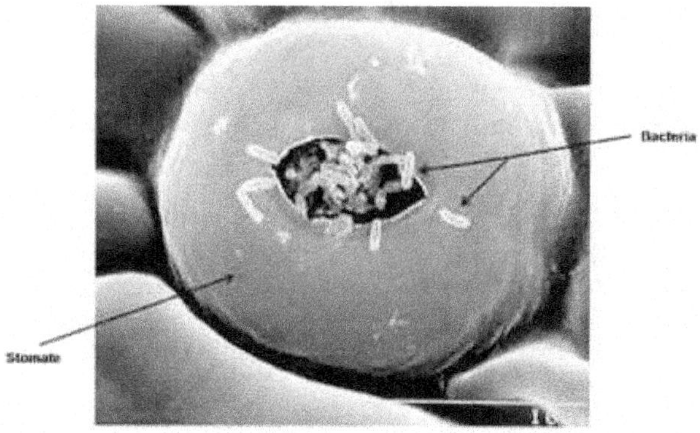

Figure 3.1: Stomate with citrus canker bacteria

After entry, the bacteria begin to multiply. As a defense mechanism, rapid plant cell divisions occurs. Canker lesions are characterized by the development of a large number of hypertrophic cells and a small number of hyperplastic cells. There is cell death resulting in epidermal rupture, initially causing a small blemish or discoloration on the leaf at the surface. The rapid cell division is called a "hypersensitive reaction" (HR). The process may be described as miniature volcano, where there is increase pressure at the surface due to increased volume created by cell enlargement and division. The raised lesion at the surface, contains necrotic plant cells. The infection process creates a polysaccharide matrix, in which the *Xac* cells reside.[14]

The bacterium is *Xanthomonas axonopodis* pv. *citri* or *Xac* for convenience. An acceptable synonym is *Xanthomonas citri* spp. *citri* or *Xcc*. Previously, it was named *Xanthomonas campestri sp. citri* or *Xcc.* The bacterium belong to the gammaproteabacteria class. *Xanthomonas* is a large genus of bacteria that cause disease in hundreds of plant hosts, including many economically important crops.[32] The bacterium is a straight, rod-shape on the order of 1 to 3 microns in size.[38]

3. Appearance of Lesions

The earliest symptoms are tiny slightly raised blister-like lesions. As the lesions age, they first turn light tan, then tan-to-brown and a water-soaked margin appears, often surrounded by a chlorotic halo.

Figure 3.2: Citrus Canker on Fruit

This water-soak margin may disappear as lesions age and may be not as prominent on resistant cultivars. With time, the center of the lesion becomes raised and spongy or corky. These raised lesions from stomatal infection may be visible on both sides of the leaf.[14] Eventually, the centers of leaf lesions become crater-like and may fall out, creating a shot-hole effect.[12, 25] The appearance can change if micro-organisms have entered the lesion, or if the entry is caused by a wound.[16]

4. Changes in Lesions with Time

Within a laboratory setting, where inoculum containing the citrus canker bacteria has been injected into a citrus fruit, the first signs of a lesions may resemble a pin prick with a diameter of one millimeter (0.039") or less.[14]

The lesions may increase in size to a maximum diameter of approximately 10 mm (0.39").[12] Based on limited experiments with Duncan grapefruit plants, it has been estimated the lesion diameter increases in size at the rate of approximately one millimeter per month for the first 6- 8 months. With time, lesions become more numerous and show a raised appearance and the halo patterns tend to coalesce. It has been reported in an article by Dr. Canteros, that most misidentification occurs when only old lesions are available.[3] Windscar, insect damage, scab, and alternaria-leaf-spot can be easily confused with canker to the untrained eye.[3] The yellow and brown halos may diminish or disappear with time, resulting in lesions which can be mistaken for citrus scab.[43]

Comparative changes in lesions appearance, from month to month, may be observable. The Department has provided images of these changes, and they are posted on the website. Judging from images of canker development, only very approximate estimates of lesion ages can be made in terms of months in the early stages of development. Pruning of infected foliage by owners and normal leaf drop can also make estimation of initial onset of the disease more uncertain.

As discussed in Chapter 1, pruning shears and insecticides are probably the first line of attack for homeowners against many diseases and pests. Where lesions are prevalent and highly visible on limbs and fruit, some homeowners may cut down their trees. As discussed in the Section 6, homeowners in study site areas could unknowingly destroy valuable epidemiology data.

5. Local Infections and Inspection Problems

The lesions are local or non-systemic in nature as the bacteria do not enter the circulatory system of the plant. Thus, each new lesion depends on successful entry and multiplication of bacteria. If a leaf with lesions were to fall off, a subsequent inspector might consider the tree to be a healthy tree. Without lesions, there is no way to know if a tree has canker, except if it were moved to a containment greenhouse for observation.

Under natural conditions, it would be expected that canker lesions and the number of infected trees within a location should become more numerous with time. However, this is not always true. In an experiment conducted in Argentina in 1992, disease incidences increased for about 4 months and then surprisingly decreased in the next four months, until there were nearly zero disease incidences at the end of the 8 month experiment.[8] This decrease was attributed to leaf drop during the colder months.[8]

Pruning can be a problem for canker detection. Homeowners may decide to prune off of diseased limbs of the citrus in hopes of curing the tree of canker or avoiding the removal of the tree by the Department. In any case, the pruning would delay discovery. Thus, conducting a survey of infected trees during an active eradication program, can lead to errors.

From October 1995 to January 1997, the program employed a "hatracking control" protocol where all branches were cut from the infected citrus tree. Also, a homeowner would naturally try to control the spread of the disease, by trimming infected branches. The success of these control efforts would likely depend on the timing, as the winter and spring months (approximately November to May) are drier and colder conditions which would limit canker transmission. The Department released no assessment on hatracking program, just the conclusion that it was not working well.

6. Conditions Affecting Natural Dispersion

Natural dispersal is the result of wind blown rain. A number of factors play a role in the release, transport and sedimentation of bacteria on the host plants, adding uncertainty to the dissemination of the disease. Water from rain or irrigation systems is a necessary element for the infection process, but other factors are important in the infection process. Long distance dissemination of citrus canker may also occur by purchasing and planting of infected citrus with canker in the asymptomatic or early stages of development. The pruning of branches with visible lesions may delay discovery.

Rain, Wind and Gravitational Forces

In low wind conditions, the rain droplets will fall nearly straight down, as a result of gravity. If new flushes at the top of the tree are infected, then the destination of most released bacteria will be the lower foliage or the ground. If citrus trees are sufficiently close, adjacent trees could also be infected. The process of rain drops splashing on foliage and locally dispersing inoculum is called "splash dispersion."

The bacteria oozes out of the lesions when in contact with water. If the infection process is to be successful, a sufficient quantity of bacteria must flow out of the lesions. The concentration of bacteria, as released in water droplets, can only diminish from the point of origin to their final destination.

Splash dispersal is a term also used to describe the mechanics of release and dissemination of fungal diseases. When rain droplets strikes the surface of a leaf, the impact can cause the dry fungal spores to be released into the air. There is no similar "tapping" release phenomena with citrus canker.

Wind velocity, wind direction and gravitational forces determine the trajectory of bacteria laden rain droplets at any point in time. Due to baffling effect within the canopy, the downstream wind velocity can be considerably lower than the ambient wind velocity. Experiments conducted by the USDA/ARS in simulated conditions are discussed in Chapter 6.

An online article indicated that a threshold wind velocity of above 8 m/s aids in the penetration of bacteria through the stomatal pores based on experimental work.[12] The original reference was from a Japanese agricultural station report in 1975 and not reviewed. A related research note by Serizawa et al. in 1981 showed an increase in citrus canker incidences with wind velocity based on measurements at three wind velocities.[27] The maximum wind velocity in these experiments was 8.5 m/s (19.0 mph). The article suggests that at higher wind velocities, leaves will turn resulting in more contact of rain with the underside of leaves. This wind velocity within a canopy would depend on other many other variables including canopy density. Further, the article indicates that high winds may cause injuries, particularly with trees with thorns in the axillae. The article concludes:

> Therefore, under the weather conditions of light rain with strong wind, the disease accelerates as a result of increase of both the amount of pathogen spreading and the number of injuries on leaves.[27]

The above discussion should include "man-made" rain, in the form of overhead irrigation systems, used both in nurseries and groves. The light and repeated sprays during windy days, would seem ideal for the natural dispersal of citrus canker in groves and nurseries. Long, heavy downpours may result in a diminishing release of bacteria, and the net result can be to wash off bacteria.

The claim that a certain threshold wind velocity exists lacks experimental scientific study. It is likely this claim was made because many of Florida's tropical storms exceed 19 mph.

Density of Plantings

Under natural conditions, canker dissemination beyond the tree's canopy would be related to the chance a raindrop laden with bacteria would come in contact another citrus host. The densely packed plants in a nursery would result in a nearly continuous field of citrus foliage, so chance of contact to adjacent plants would be almost certain. At the other extreme, residential areas are sparsely populated and this fragmented pattern of citrus, resulting in lower chance of contact.

Residential Areas: The density of citrus within residential areas is estimated to be 1,957 citrus trees/square mile (3.13 trees/acre) based on the Florida field study.[11] An article by Schubert et al. 2001 states a square mile section contains approximately 2,000 parcels, and about 50% of the parcels contain citrus.[25] Those parcels containing citrus have 2 to 3 citrus trees, or a range would be 2,000 to 3,000 citrus/square mile or 3.13 to 4.69 trees/acre, so the field study estimate of 3.13 trees/acre is at the low end of the range given by the Schubert et al's article.

Groves: Planting spacing for citrus can vary considerably. A 15 x 25 ft spacing (25 ft between rows) results in 116 citrus/acre, while 10 x 15 ft spacing results in 290 citrus/acre. The UF Extension Report indicated the general trend has been to plant at higher density given the decline in available acreage.[40] Based on the 2012 CCEP Comprehensive Report, the average planting density in the eradicated acreage was 129 citrus/acre.[6]

Nurseries: The planting density of citrus can vary widely. The 50 acre A. Duda and Sons nursery was destroyed during Canker War II, resulting in the destruction of 500,000 citrus, or approximately 10,000 citrus/acre.[21] Also, it was reported that a second nursery, the 8 acre Sweet's Citrus nursery, destroyed during Canker War II, had 200,000 citrus or approximately 25,000 citrus/acre.[21] For a nursery density of 10,000 citrus/acre, and grove density of 129 citrus/acre, the nursery density is 77 times more dense than a grove.

The ability for canker to be disseminated should be related to the ability of rain to come in contact with other citrus plants. If all other factors are the same, canker should be most infectious in nurseries, less so in the groves and the least in residential areas.

In summary, the approximate estimates of planting densities are:

Residential areas:	3.12 trees per acre with high variability.
Groves:	116 to 290 trees per acre, CCEP average = 129 trees/acre
Nurseries:	10,000 to 25,000 trees per acre, based on very sparse data

The mean distance between trees, at a density of 3 trees per acre, would be approximately 120 ft if a random distribution is assumed. It cannot be presumed that residential trees are randomly distributed, and a modal distance, or most likely distance between trees would be more useful in epidemiology studies. The FDACS/DPI likely has one of the most extensive databases on the density of citrus in these three environments as inspectors during the 10 year period obtained data on both healthy and infected trees. Someday, it is hoped this information will be shared with the public.

Climate and Seasonality

Warm weather is another contributing factor to the infection process as higher concentrations of bacteria are produced during the summer months. Generally, the months from November to March in Florida have lower temperatures and less rain, conditions less amenable to bacteria production.

Cold weather in the winter months can delay the appearance of canker. This is called "overwintering." Citrus canker overwinters in lesions in autumn on diseased leaves, twigs and stems. In the summer, the lesions on leaves are the primary sources of inoculum for fruit infection.[18]

Also, the court decision of Denny (*Denny* v. *Conner*, 463 So 2d 534, Fla 1 DCA) more eloquently put the same concept, "It may lay dormant in apparently healthy tree for some months (one botanist opined up to 18 months) after exposure to infected plants before manifesting signs of the disease."

Similar statements concerning inactive periods are presented in an article by Pruvost et al., "Thus there is a discontinuity in the *X. axonopodis* pv. *citri* lifecycle under such environmental conditions where canker is not active, (low bacterial population, no bacterial multiplication, lack of new susceptible host tissues) ."[22]

The new flushes of a citrus tree, which are more susceptible to canker, may occur several times a year, in the spring, summer and fall, and less likely in the winter. The slow progression from leaf to fruit is supportive of defoliation as a means of control, as noted by Dr. Civerolo of the USDA/ARS points out, "Removal of overwintering inoculum sources by pruning infected late summer and autumn shoots effectively reduces infection the following spring."[18] Links to articles related to climate and the spread of citrus canker are provided on the website.

Host Species/Cultivars Susceptibility

Members of the Rutaceae family are potential members of citrus canker. There are 160 genera and over 1600 species in the Rutaceae family. As new hybrids are developed, the list of citrus cultivars increases. Thus, it is impossible to test every species for citrus canker. Also, the susceptibility of each of these species may vary depending on the strain of the bacterium.

FDACS/DPI developed a list of the most common cultivars in Florida. The list below covers only some of the more common citrus.[25]

Highly susceptible: *Citrus x paradisi*, (grapefruit), *C. aurantifolia*, (key/Mexican lime), *Ponicirus trifoliata* (trifoliate citrus and hybrids, *C. sinensis* (sweet orange cultivars: Hamlin, Naval and Pineapple).

Moderately susceptible: *C.sinensis* (sweet orange except Hamlin, Naval and Pineapple), *C. aurantium* (sour orange), *C. limon* (lemon, but some cultivars are more susceptible than others), *C. x tangelo* (Orlando tangelo), and *C. maxima* (pummelo, some hybrids are less susceptible than the species).

Susceptible: *C. reticulata* (Mandarin, tangerine), *C. aurantifolia* (Swingle, Persian/Tahiti lime).

Resistant: *C. medica* (citron, although this species gets severe stem lesions), X *citrofortunella microcarpa* (calamondin), *C. japonica* (kumquat).

However there were questions regarding other ornamental plants, which residents could buy locally and were in the Rutaceae family. A book published by Swings and Civerolo in 1993, listed two ornamentals and members of the Rutatceae family, orange jasmine and wild lime as host plants of citrus canker.[32]

Orange jasmine (*Murraya spp.*) is a common hedge plant in Florida. The Department responded it had not been shown to be a host in the field.[5] However, the Department did not state if laboratory testing had been conducted. The orange jasmine was been determined to be a host plant for citrus greening and at present, is not sold Florida. When cut down, orange jasmine tends to grow back easily unless roots are removed. Since it is a commonly used in hedges in the landscape, it is likely a serious problem in the control of citrus greening.

Wild lime (*Zanthoxylum fragara*), is a native Florida plant, and a larval host to the giant swallowtail (*Papillo cresphontes*) and the Schaus' swallowtail (*Papillo aristodemos*). The Schaus' swallowtail is known to exist only in the Miami/Homestead area and is highly endangered. Wild lime was not inspected nor eradicated during the eradication program.

Knowing the susceptibility of the cultivars has led scientists to investigate the cellular defenses which these less resistant cultivars have against intruding *Xac* cells.[19] This research might lead to new cultivars, which have better internal defenses.

Threshold Concentration Levels

A threshold concentration of bacteria, as measured in colony forming units per ml (CFU/ml), is the concentration required at entry to cause an infection. Under favorable environmental conditions, the minimum level of *Xac* is 10^4 to 10^5 CFU/ml if entry is through the stomata cells, and 10^2 to 10^3 CFU/ml if entry is through wounds in the plant.[23] The threshold's lower limit may be considered a likely limit to cause infection in field conditions.

An article by Drs. Graham and Gottwald[45] suggested it is possible that a single bacterial cell could cause a lesion to form. These infiltration experiments were under special laboratory conditions of high impact pressures. At the International Citrus Canker Research Workshop on June 20, 2000, Dr. Schubert was cautious on extending these results to real world circumstances and indicated he would like to see more done into this area.[46]

Rain water which is free of bacteria may dilute or wash off canker-laden rain water on the recipient host. Heavy afternoon rains with little wind, are likely to promote infections more within trees with established canker infections than more distant locations. The concentration of bacteria released from a lesion tends to diminish with time. The bacterium may survive in landfills or on other plants, but it does not multiply. It is not saprophytic in these environments.[34, 35] On hard dry surfaces, the bacteria die quickly. Sunlight UV rays help destroy the bacteria that has been released in water.[12] Most bacteria on hard surfaces die within 48 hours.[34,35]

7. Citrus Leafminer

Wounds in citrus trees have long been identified as increasing the susceptibility of a citrus tree to canker. Wounds caused by high winds can not be avoided, however there are control measures available for insect wounds.

The citrus leafminer (CLM) is a moth whose larvae create serpentine mines within the leaves of a citrus tree making the foliage more susceptible to citrus canker. It was first discovered in Florida in 1993 and quickly became established in many counties in Florida.[15] Citrus leafminer distinctive mines are easily identified on the leaves. The larvae of other insects create wounds in citrus. Citrus is a host plant to the larvae of the giant swallowtail (*Papillo cresphontes*). Gardeners plant citrus and other Rutaceae plants so they can attract swallowtails in their yards.

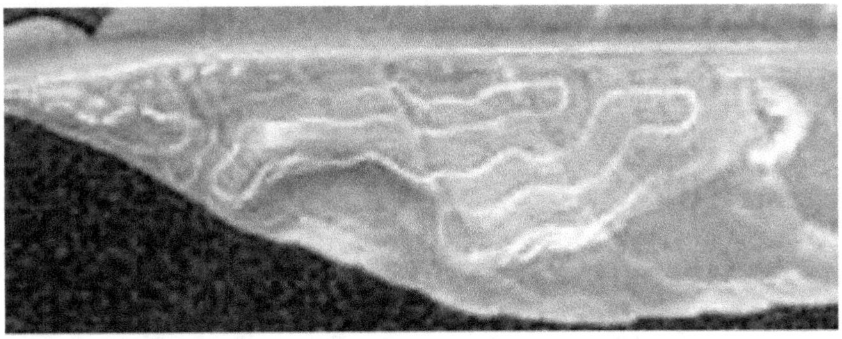

Figure 3.3: Mines made by citrus leafminer.

Studies have shown the rate of infection by citrus canker with leafminer damage is higher than undamaged leaves. This was attributable to the hyper-susceptible tissue exposed by the leaf miner galleries. [12]

Not all researchers voiced the same level of concern of the wounds made by CLM. Dr. Canteros, an expert scientist in citrus canker noted that the first canker lesions on canker infected plants do not appear in mines in groves in Argentina.[3] This would seem to indicate that CLM may not make citrus canker disseminate more rapidly by windblown rain, but could make the disease more severe on an infected tree over time.

As grove owners have become aware of the harmful effects, it is likely they take control measures. There are a number of means of controlling citrus leafminer. The University of California/Integrated Pest Management website, lists general five control methods: traps, biological control, cultural control, physical control and chemical control. Traps and biocontrol are two environmentally safe alternatives. Biological control consists of release of parasitic wasps, a natural predator of citrus leafminer.

Dr. Timmer a researcher at UF/IFAS reports, "With effective bio-control agents and specialized products to control outbreaks, CLM no longer represents a disaster for the Florida citrus industry. Maintaining effective levels of bio-control, however, requires avoidance of broad-spectrum insecticides which are toxic to parasites. This factor emphasizes the need for consideration of all pests in decisions concerning pesticides to be applied."[33] Grove owners may prefer chemical control, as they can treat for a number of insect problems.

Residents are likely not to react as quickly and completely as commercial growers. However , access to traps is not a problem today. CLM traps are now sold over the internet and at commercial outlets in Florida for residents.

Research has shown that the adult citrus leafminer is not an efficient carrier (vector) of canker bacteria.[1] The female CLM is likely a frequent visitor of citrus, as she looks for the appropriate host plant to lay her eggs. By extension, this research reduces the likelihood of other frequent visitors in the insect class as being effective vectors of bacteria.

8. Unsupported Mechanisms of Dissemination

The discussion in this section is rarely found in other articles on citrus canker, but should receive more attention. Science relies conjecture, because this first step in the progress of science. However the next step is investigation and whenever possible, repeat control experimentation, to establish a scientific basis of new theories. The mechanisms or means of dissemination discussed in this section, were promoted by the Department, but appear to be only weakly supported by experience and unsupported by experimentation.

This lack of experimentation is not based on lack of facilities. The USDA/ARS in their various containment greenhouses in Ft. Pierce, Florida, Beltsville, MD and other locations have the capability to conduct repeated tests on the means by which citrus canker can be transmitted.

In 2007, the USDA/APHIS did a thorough risk analysis of canker dissemination resulting from the shipment of asymptomatic citrus fruit, from areas where canker had been discovered.[38] The risk of dissemination was determined to be extremely low. This study was released only after the eradication program ended.

Birds, Cows, Horses and Insects

The Australian Public Broadcast in an August 27, 2004 announced to their listeners and posted on their website, "it can also be spread on the feet of birds that land on diseased plants." Where was the Australians getting their information? Perhaps they were just echoing information on the USDA/APHIS website. Or perhaps it came from the Florida Department of Agriculture website.

From the International Citrus Canker Research Workshop transcript as posted on the FDACS/DPI website, Dr. Gottwald, page 74, "... there is potential for spread of bacterium from surface materials. Here it is on the bark of citrus. Feathers. We are looking at the potential for birds to spread it around. We have seen in Argentina were it gets on the fur of a horse or cow and it could be transmitted or move down rows and spread down the foliage rows."[22] If a cows or horses can transport citrus canker, this opens the door to a whole host of mammals and reptiles that could come in contact with citrus fruit, such as cats and dogs, hence disseminate the disease.

Similarly, in 1997, Dr. Gottwald wrote, "in addition *[to inadvertent human transport],* it is possible that animals such as birds and insects can passively carry *Xac* bacteria from tree to tree. "[9] There is no experimental research to show birds can disseminate citrus canker

In a 1992 article by Dr. Gottwald, an abandoned bird's nest was discovered in a lemon tree which had canker lesions. Another citrus tree with canker lesions was discovered 320 m away. Thus, it was suggested birds could have disseminated citrus canker.[8] However, there is no reference to any experimental studies to support this means of dissemination.

Drs. Jetter and Civerolo (Chief Plant Pathologist, USDA/ARS) wrote in a paper in 2001, "Long distance dispersal of Xcc by animals, birds and insects has not been conclusively demonstrated." [18] Citrus has many insect pests, including aphids, grasshoppers, mites, and citrus leafminer.[17] Through scientific study, the adult citrus leafminer was demonstrated to be not an efficient carrier of Asian citrus canker.[1] Since this insect is obviously in frequent contact with the citrus plants, and does not vector *Xac*, by extension, other insects and reptiles, such as lizards, would be less likely to vector *Xac*.

The FDACS initially included the suggestion that animals, birds and insects were potential dispersers of citrus canker bacteria on their citrus canker website in year 2000, but later removed this reference. It was an odd statement, as if birds and insects were not animals.

Tornadoes

A 1997 article indicated that a tornado pass through a residential area of Miami on January 3, 1996. The authors claim to determine from examination of lesion ages of new infections, the tornado disseminate citrus canker 6 to 7 miles.[9] However, the article did not provide any data to demonstrate how this distance was determined. Citrus canker requires wet conditions for release of bacteria. There is no experimental data which shows canker bacteria can be released and infect another plant by wind alone. High winds can do damage to citrus trees, and aid in bacteria entry, which is one component in the infection process.

The claim was repeated by Dr. Gottwald in a presentation in November 2000 in the Broward courthouse. It is likely this same viewgraph was presented at the International Citrus Canker Research Workshop in June 2000. A short note on the tornadoes and hurricanes is posted on the online supporting documents website.

Postal Workers, Meter Readers and Sanitation Trucks

The FDACS posted on their website from 2001 to 2006, "Citrus canker is highly contagious and can spread rapidly by windblown rain, people carrying the infections on their hands, clothing or equipment and by moving infected or exposed plants or plant parts." In an article by Gottwald et al., the authors reported, "Driving forces of an epidemic… Sanitation department trucks passing through alleyways brushed up against infected trees and spread infections to nearby trees along alleyways."[9] Citrus trees along alleyways? While lacking in specifics, it was consistent with the Department's theme that citrus canker was highly contagious.

By extension, within residential areas, anyone which would regularly come on to residential properties in their official capacity, was thought to be contributing to the dissemination of citrus canker. This could include postal workers and meter readers. There was no experimental studies to show the regular visits to residential properties would disseminate citrus canker.

Workers' Hands, Clothing, Pruning Tools and Packing Crates

The UF/IFAS Citrus Pest Management Guide states, "Workers can carry bacteria from one location to another on hands, clothes, and equipment. Grove equipment can spread the bacteria within and among plantings, especially when trees are wet. Workers can carry bacteria from one location to another on hands, clothes, and equipment. Grove equipment can spread the bacteria within and among plantings, especially when trees are wet."[36]

Their comments are directed to workers in groves, when the citrus trees are wet. This is quite different for the Department's comment of "people carrying the infections on their hands, clothing or equipment" causing potential spread of disease as posted to the Department's website.

The question of importance may be whether this is handling of fruit is a significant pathway in the large scale spread of the disease. Under dry conditions, there is no scientific evidence that the disease can be transmitted on someone's hands. It is also apparent that sanitation procedures need to be higher in the groves, where workers may handle hundreds of citrus trees every day. So, while there scientific evidence may be limited on the spread of canker by pruning equipment or packing crates, sanitation measures may be taken until otherwise shown to be unnecessary.

The USDA/APHIS did a comprehensive study to determine if citrus fruit from groves in areas where citrus canker was known to exist, could pose a risk of disseminating citrus canker. The study determined the risk was very low.[37]

Dissemination from Soil, Leaf Litter and Weeds

Weeds as hosts and soil or leaf litter as environments which prolongs the longevity of the bacteria have been discussed in the literature, but there never seemed to be an effective mechanism to disperse the bacterium from the soil to citrus trees. Further, there does not seem to be experimental data to back up these claims. Disinfection of lawn mowers was part of the Department's regulation in areas where citrus canker was discovered. However, it seemed difficult to enforce.

Dr. Gottwald in an online publication wrote in 2002, "However, there is to date no direct evidence that *Xac* surviving in low numbers on weed hosts or in the soil can serve as sources of inoculum for epidemic development."[12]

9. Canker Hosts and the Two Strains Problem

Using DNA analysis, scientists determined two important strain subgroups of Asian citrus canker bacterium in Florida, the "Manatee" and "Miami" subgroups. These two subgroups have the same host range, which includes most of the common citrus cultivars. So, for the eradication program, it was unnecessary to determine the subgroup.

However, a new subgroup was discovered in Palm Beach in May 2000. This new subgroup was called the Wellington strain and could infect only key limes and alemows (a species of citrus rarely grown in Florida). This new subgroup was first revealed in a presentation by Dr. Sun with the FDACS/DPI on June 20, 2000, during the International Citrus Canker Research Workshop.[10] It is likely the discovery probably took place months earlier, but was finally confirmed as an unique strain in May 2000.

The lesions created by these three subgroups are indistinguishable. If lesions were discovered on a key lime, only DNA testing could tell if the canker was Wellington or the other two subgroups. Through genetic analysis, some tests were able to differentiate these strains, many of the genetic markers were identical. Therefore, DNA testing may not always provide definitive results.[10, 11] Complex and possibly ambiguous DNA work was never needed to tell these two strains apart. In fact, it was quite simple and the same method used in 80 years ago. The lesions were cultured and injected into a grapefruit. If they developed lesions, it was not the Wellington strain. Sun in a published article states, "The simplest and most reliable method used to distinguish two strains is to test their pathogenicity on two grapefruit plants. "[11]

Recommended Policy Changes Ignored

The article by Sun et al.[31] further states, "Based on preliminary results, a group of plant pathologists from Florida recommended in February 2001 that the Wellington group of strains be characterize as a unique group of strains of *X. axonopodis* pv. *citri* and that all of its host plants, Key/Mexican lime and alemow be removed within 579 m (1,900 feet) from a diseased plant."

No procedures or rules were ever implemented to safeguard against cutting trees that were not hosts to the Wellington strain of canker. A new law was passed by the legislature requiring the destruction of all citrus regardless of the strain which was present.

Also, it is suspected very little was done to identify whether the infected lime trees in Florida had the Wellington strain. Sun states that in a six month study in a 54 square mile area of Broward County, they sampled only <u>eight</u> infected lime trees. All eight lime trees tested positive for *Xac-A*, ruling out that they were infected with Wellington. However, why test only eight trees? No special DNA testing is required, only a pathogenicity is needed. It is possible that FDACS really did not want to know the extent of the Wellington strain.

FDACS produced a map showing the locations where the Wellington variety was present. It is confined solely to a small area within Palm Beach County. This is very likely the result of lack of investigation of its presence as it creates a regulatory complications. The map has been posted to the website on the Biology page.

There is no mention of any investigation of the presence of the Wellington strain in the Homestead area of Miami-Dade county, where approximately 3,000 acres of commercial lime groves existed until Canker War III began. The is possible that by the time the Wellington strain had been confirmed, a majority of these groves had already been destroyed.

10. Discovery of Citrus Canker in Residential Areas

At the June 2000 International Citrus Conference Research Workshop, Dr. Schubert stated that when citrus canker is discovered in a new location, one is looking at history.[22] The bacteria may have already disseminated far from the discovery, but citrus trees are not yet showing symptoms. Once an infected tree and the surrounding trees are destroyed, then any map showing the "spread" of citrus canker in terms of eradication circles is really showing where citrus canker does not exist.

The disease has been called elusive because a positive identification of canker may not be apparent for months or even years after the initial infection. The care given to residential trees varies greatly. There may be multiple diseases and pests, and nutritional problems, which compound the identification of early detection of citrus canker.

In an interview with FDACS/DPI Director, Richard Gaskalla, stated that samples of citrus canker which are taken back to the laboratory usually confirm citrus canker.[7] He states, "Field crews are 90% accurate in their canker identification."[7] It was not possible to know how often laboratory examination were done. Inspectors generally did not arrive at homes with equipment for taking samples such as baggies, ladders and poles. The Department has not produced any report showing the frequency of inspection identification errors.

Finding the symptoms of citrus canker in residential neighborhoods can be difficult. The most susceptible parts of a citrus tree are the new flushes appearing in the spring. Unfortunately, the new foliage appears often high in the canopy of the trees, making inspections difficult. Inspections in residential areas could be done until 7:00 pm at night. Twilight conditions likely result in false negatives. Citrus trees can grow to over 30 ft high. As trees are often planted in the backyard area, the canopies of other trees are commingled. The impact of survey errors on eradication efforts is discussed further in Short Note 3.3.

In comparison to residential areas, nurseries and commercial outlets are much easier to inspect. Inspections of groves are also more rapid using pick up trucks to go down rows in the groves. Citrus trees are normally topped off, to increase yield and make harvesting easier. This likely makes inspections easier, but may make early discovery more difficult. False negatives would be the result.

11. Identification Problems: Bacterial Spot, Greasy Spot, Melanose and Citrus Scab

Several diseases produce symptoms that look similar to citrus canker particularly in the early stages. One of these diseases is citrus bacterial spot. It had been diagnosed in 1984 as a form of citrus canker. Other similar fungal diseases include greasy spot, melanose, citrus scab and alternaria. In fact, when citrus canker was first discovered, scientists thought to be a form of citrus scab, a fungal disease.

In 1981, citrus canker was identified in Mexico. Scientists considered this a new strain of canker, called D-strain canker at the time. The symptoms fit citrus canker with small, raised watersoaked pustules surrounded by chlorotic halos on succulent leaves and twigs. Initial testing by the USDA erroneously confirmed that this was a new strain of citrus canker. Based on laboratory analysis, quarantines against Mexican citrus were implemented by the USDA/APHIS. Further testing showed the disease was not caused by bacteria, but was a fungus.[29]

The identification problem can not be well demonstrated without color images. For this reason, this topic is expanded in the online supporting documents website, with color images showing the similarity of symptoms of CBS, greasy spot and citrus scab on citrus leaves.

Not only can citrus canker be incorrectly identified, but an article in the Miami-Herald, in June 2000, citing many incorrect identifications of citrus trees. Other trees, such as mangos, and avocados, which were identified as citrus trees and reported to the CCEP hotline.[20] It was surprising because mangos and avocados leaves are generally very different from citrus trees.

During the CCEP, the general policy of the Department was to have an initial inspection of the property to find potentially infected trees. If there was some doubt, this would be resolved in a follow up inspection done by a train plant pathologist. This procedure was not mandated by Department rules.

Very often, the only identification in the eradication program was by visual inspection. A list of possible laboratory tests to determine reliable identification of citrus canker was prepared by the Department in response to residents questions at a public hearing in November 2001 and is posted on the supporting documents website.[5]

According to Dr. Canteros in a 2004 article, proper identification of citrus canker requires laboratory testing:[3]:

> Most misidentification occurs when only old lesions are available. Wind scar, insect damage, scab and alternaria leaf spot can be easily confused with canker to the untrained eye. ... The use of more than one highly desirable to avoid the occurrence of false positives and false negatives. Serological and molecular methods can be used together with pathogenicity test.

In older lesions, the yellow and brown halos may diminish or disappear entirely, making identification more difficult.[43] (image provided on website by Paul Chaloux, USDA/APHIS)

There is considerable new research to improve in detection protocols and test procedures. Brazilian scientists attribute much of their success of their eradication program to intense and timely grove and nursery inspections.[44]

Citrus Canker Identification Problem Results in a Lawsuit

One incorrect determination could potentially result in the illegal destruction of hundreds of healthy trees in a residential setting by the Department. The canker law gave the Department complete control over the determination of which trees were infected with canker. All that was necessary was one small lesion on one leaf. The identification problems became subject of the continuing lawsuit against the Department (See Broward Case 3 in Chapter 5).

During the trial, the laboratory technicians testified that the scalpels used to extract the lesions were not routinely decontaminated, so cross contamination was possible. Also, the existence of both the Wellington strain, and the similar looking disease, citrus bacterial spot, complicated laboratory analysis.

On July 23, 2003, Judge Fleet ordered the Department to perform all tests to reliably diagnose and determine if a subject tree is infected with Asian strain citrus canker. This testing included the hypersensitivity reaction test (pathogenicity test). The FDACS appealed the decision and the appellate court overturned the Broward Court decision. The Commissioner of Agriculture, Charles Bronson, stated publicly that the Broward Court was trying to micro-manage the program.

12. Citrus Canker as a Plant Health and Economic Issues

The degree to which citrus canker harms the plant in terms of reducing its yield or limiting its longevity has been a contentious issue. Early fruit drop and defoliation have been observed. [12] However, productivity losses have been reported as low.[3] Citrus fruit with lesions is not marketable, so grove owners lose value even if the yield does not change. The juice of the infected fruit however is generally unaffected, but the profit margins for juice are less. Control methods, including windbreaks and additional copper spraying, increase operational costs. However, any discussion on health deficiencies is complicated by the fact, that residential citrus trees are generally not grown for maximum yield. Other factors impact yield including inadequate cultural practices (fertilization, irrigation, pruning, cold protection and rehabilitation), and pest disease and weed management.[17]

Prior to the eradication program, estimates of a potential loss in yield to groves with canker, were for the most part, theoretically, as commercial grove owners would normally remove infected trees soon after discovery. Plant pathologists opined in the media that yields could drop from 5 to 50% if a citrus tree was infected with canker. Now that Florida is living with citrus canker, the impact should be seen in statistics on increases in culled fruit with citrus canker lesions from packinghouses. Unfortunately, no post eradication studies have been conducted by regulatory agencies, such as USDA/APHIS.

The Department supported the statement that citrus canker is a devastating disease by reference to a published study in the Maldives Islands, where there was a decline in lime trees which were heavily infected with citrus canker.[24] Only Key limes are grown there — one of the most susceptible species to canker. Trees were grown from seed, not grafted as in Florida. The soil is highly alkaline, so this is the only species that can successfully be grown. The conditions are ideal for canker dispersal, with year round warm and windy conditions. The soil is nutrient poor, and a lack of care and pest control were common problems. Interestingly, some of the islands did not have canker. Of 115 islands reporting the status of their trees, 35 islands did not report any with canker.[24]

The Maldives Islands experience is not easily extended to Florida's situation. The Department could not show any case of trees dying in the US due to canker. It is ironic that the only species which the Department could claim was killed by canker is Key lime and it is no longer commercially grown in South Florida. The loss of approximately four square miles of Key lime groves can be attributed to the eradication program and grove owners who voluntarily converted the groves to other crops or sold their land to developers.

13. Control Measures

Until 2006, the CCEP's objective was effective eradication through destroying both infected and healthy trees. There was opposition to the approach from both residents and plant pathologists. Other scientists argued that a more economically sound approach was "living with canker" consistent with control approach taken by Argentina. In Florida's case, the USDA determined in late 2005, after considerable discussion with expert scientists and epidemiologists, that eradication is infeasible.

The University of Florida/Citrus Research Center publishes a Pest Management Guide (SP 043) which identifies control methods such as planting disease resistant cultivars, copper spraying, control of citrus leafminer and windbreaks. It is typically updated every year. The most recent guide strongly supports windbreaks as the single most effective means of dealing with citrus canker.

The Citrus Health Response Program, a cooperative program between the USDA and FDACS. and supported by efforts of the University of Florida, provides a framework to actively manage and protect citrus from diseases including citrus greening and citrus canker. The program was initiated after the CCEP ended in 2006.

14. Concluding Remarks

It was not possible to cover all aspects of citrus canker biology, so only selected topics have been presented in this chapter. The basic infection process has been well identified by plant pathologists. Research continues to explore the fundamentals of plant susceptibility/resistance and related plant-microbe interactions. Many of the biological aspects of citrus canker can be found in online in the supporting documents website and at American Phytopathological Society website (www.apsnet.org).

The focus in this chapter was aspects of citrus canker disease, as they relate to the Florida epidemic and eradication efforts. Three practical aspects of the Florida outbreak, which are not frequently noted in published articles, are:

- The dispersion of bacteria by windblown rain is dependent on many factors, one of which is the planting density of citrus. From limited data, it is estimated that a nursery is about 77 times more dense than a grove and 3200 times more dense than residential areas.
- The "animals, insects and birds" as potential dispersers was never scientifically determined. Only one study investigated the potential for an insect (citrus leafminer moth) to vector canker in 2005, and came up with negative results. Neither the Florida Department of Agriculture nor the USDA/ARS Tropical Research Group in Florida participated in this study.
- The "two strain problem" required much more thorough testing of citrus canker and correct identification of the healthy trees. The Department did little to assess the degree to which the Wellington strain was present in Florida.

One of legacies of prior eradication efforts was canker was capable of spreading long distances in very short time periods following rain storms. Citrus canker was discovered in many of the gulf coast states in the period 1913 to 1915. It was considered a highly contagious disease based on various natural means of dissemination including transport by windblown rain, birds and insects. Later investigation of the dissemination of canker during Canker War I, revealed the origins of the multi-state presence was due primarily to contaminated nursery plants shipped from Texas.

Unfortunately, this "highly contagious and devastating" label remained, as it gave priority to the eradication of the disease. It was helpful in the politics of obtaining funding. This description is difficult to refute, as there are no rules dictating on what is contagious verses highly contagious. Also, the Asian strain was considered by many researchers to be more contagious than other strains. This label was convenient in banning Argentine citrus in the 1980's. However, it also had a negative side, as it could trigger trade restrictions on the export of Florida's fresh fruit.

Numerous mechanisms or means of dissemination have been suggested over time without scientific study. If the bacteria could reside in the fur of cows roaming through groves, then why not in the hair or clothing of residents. Perhaps for epidemiological purposes, the focus should be on the most likely means of dissemination of canker to new areas. A single seedling with citrus canker in a nursery, has the potential of infecting other plants within the nursery and eventually disseminating canker to many new locations.

Short Notes on the Website:

SN 3.1 The Case of the Dirty Scalpel
SN 3.2 Long Distance Transport of Citrus Canker by Hurricanes and Tornadoes
SN 3.3 Survey Errors and the Impact on Eradication Program
SN 6.3 Comparison of Department Website Justification Statements
SN 6.4 1990 Highlands Observational Study

Definitions

Hypertrophy — the increase in the volume of a tissue due to the enlargement of its component cells (Wikipedia).
Hyperplasia — the cells remain approximately the same size, but increase in number. (Wikipedia)
Pustules — Small raised areas.
Rutaceae — Biological family which includes all citrus plants.

Conversions

1 m/s = 2.237 miles/hr
1 acre = 43,560 ft^2
1 sq mile = 640 acres
1 meter = 3.28 ft

4. Opposition to the Eradication Program

> Nothing strengthens authority so much as silence.

Leonardo de Vinci

> Dear Mr. President [Clinton],
>
> The entire US lime industry is presently at risk due to actions of the governments of the State of Florida and the United States. We respectfully request your assistance in this emergency situation.

Letter from Brooks Tropicals, LLC, Limeco, Sapp Farms, Grove Services and Acosta Farms, July 18, 2000, protesting the citrus canker program.

---

1. Surprise, Surprise, Your Trees were Destroyed Today

As chainsaws invaded the peace and tranquility of the homeowners in Miami-Dade and Broward County in the summer of year 2000, many residents became enraged as the privately contracted crews went door-to-door, destroying every citrus tree they found. Had their leaders in Tallahassee gone crazy?

Often these attacks occurred without warning. One media story told of a homeowner who saw the contractor's trucks pull up to his yard as he was going to work, and assumed it was the electric company which routinely trims branches over the power lines. When he returned home after work, where his citrus trees had been growing was a pile of sawdust.

The Florida Department of Agriculture and Consumer Services Department (FDACS) were supposed to leave notices on the front doors, informing the residents that their trees were about to be cut down. Yet, prior to November 2000, these notices ("Immediate Final Orders" or IFO's) did not specifically tell homeowners that their trees were deemed either infected or exposed and would be destroyed. The IFO was simply a statement of policy as it stated that the Department was removing trees showing infection or exposed to the disease because they were within 1900-ft of an infected tree. The notice gave a resident five days to file an injunction to stop cutting.

Prior to November 2000, the IFO's were written in only in English. When the Department was forced to defend their program in court, this quickly changed. The Department developed new IFO's were written in English, Spanish and Creole. Technically, residents could contest the IFO's, but only a lawyer would know how navigate through the legal system. The process would be costly and complicated.

In general, the first time residents would know their trees had been condemned was when contractors came with chain saws. If a resident refused entry to the property, the contractor could call the police.

Few residents knew the long history of citrus canker. Accompanying each of the past eradication programs, there had been stiff opposition. However in the summer of 2000, many residents were learning a lot more about citrus canker eradication. Many horticulturists doubted the eradication program could succeed, as they felt citrus canker had become too prevalent or endemic to Florida.

The Department could have, at the onset of the 1900-ft rule, told residents they planned to cut down both infected trees and healthy trees, but certainly the Department's legal department would have cautioned against this form of honesty, particularly given the class action lawsuit that began with the 125-ft policy (Valera case). Instead, only infected and "exposed" trees would be destroyed. A letter from Commissioner Crawford, soliciting residents' cooperation in the eradication program in year 2000, (Figure 4.1) was posted on the Department's website, and likely provided at town meetings in South Florida. It implies that "exposed" trees are being discovered on properties, when in fact, it was impossible to inspect a tree for canker exposure.

The eradication program (2000 to 2006) coincided with a phenomenal housing boom, consumer borrowing on all levels, low unemployment and a roaring economy. Thus, the theme as reported in the newspapers, "Groves to Condos" seems supported by many accounts. It was reported that many of the grove owners were also in the real estate business.

Controversy of the eradication program extends to numerous political, economic and environmental aspects which are beyond the scope of this book. However, numerous links are provided in the supporting documents website with opinions both in favor and opposed to the 1900-ft policy.

The initial eradication of healthy citrus trees under the 125-ft policy began in January 1997 and lasted until the moratorium on cutting began in February 1998. The moratorium lasted until June 1999, after which the eradication program resumed. Prior efforts to control canker were to strip the infected trees of all limbs (hatrack), so hopefully, the trees would grow back without canker. The urban areas near the Miami International Airport were the focus of eradication efforts during the early years. Later efforts included areas further north and south of the airport and eventually extended into Broward County.

During this time, the 125-ft circles (1.1 acres) would likely result in destruction of the infected tree, and healthy trees in yards adjacent to the culprit infected trees. The 1900-ft rule meant all citrus within 260 acres would be destroyed. Since yards in Miami-Dade and Broward counties are ¼ acre or less, an eradication circle could contain 800 yards. The eradication circles would tend to overlap, so a large swath of Miami-Dade and Broward counties would be clear cut of all citrus.

CITRUS CANKER

The threat to Florida agriculture

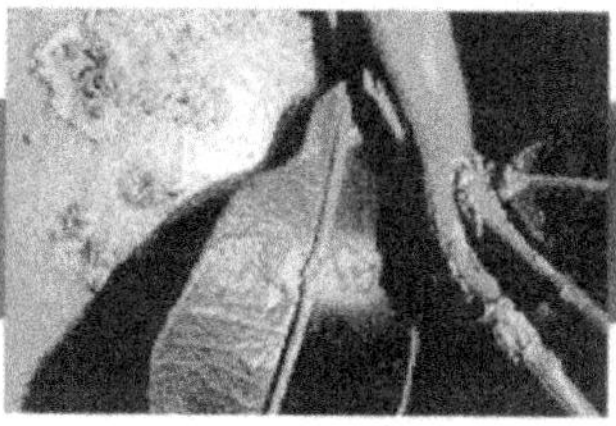

space

Message from Commissioner Bob Crawford

Dear Friends:

The Department of Agriculture is currently waging a battle against one of the most devastating, highly contagious diseases known to citrus trees, and we need your help.

Citrus Canker causes trees to weaken, lose leaves and drop fruit prematurely. Infected fruit will have visible canker sores which impact the quality of the fruit. Eventually, diseased trees will produce a small, substandard crop. There is no cure for Citrus Canker. The only known eradication method is cutting down and disposing of infected or exposed trees.

Inspectors with our department are inspecting citrus throughout the state. You may have already been visited and found to have infected or exposed trees on your property, which have been or will be cut.

I appreciate the sacrifices you are making as part of the effort to put an end to this serious problem. I am committed to trying to save the millions of backyard fruit trees throughout the state that have not yet been impacted by canker as well as our $8.5 billion commercial citrus industry.

Your assistance and cooperation are helping us to speed up the eradication effort. This is important, because once canker has been eradicated and the quarantine lifted, residents can replant citrus trees as soon as federal authorities allow. Those new trees will bear healthy, fresh fruit for you, your children and grandchildren to enjoy.

Sincerely,

Figure 4.1 Letter from Commissioner Crawford

2. Opposition to the Program with 125-ft policy

Residents within Miami-Dade County were very upset when cutting crews came to either hatracked or removed their citrus trees. However, there were no organized groups formed to oppose the eradication program. The responsible party to this destruction, the Commissioner of Agriculture, was nearly 500 miles away in the capital city of Tallahassee, Florida. So, beyond sending an angry letter to Tallahassee, Florida, there was no effective means to protest the program.

Dr. Graham, professor of plant pathology at University of Florida/ IFAS wrote in 1998[1]:

> The eradication agencies are even more hampered than ever by groups of residential property owners who have legally impeded access to their property for survey and successfully lobbied for a moratorium on destruction of 'exposed trees.'

However, it is doubtful that there ever was a successful lobby by residents for a change of policy as claimed by Dr. Graham. Outside of the class-action lawsuit, there was no organized group of residents opposing the program. Dr. Graham is incorrect to suggest residents could legally impede access of properties prior to year 2000. The normal procedure taken against residents who blocked entry to their property was to call the police. Before year 2000, the issue of the "police powers" to enter, inspect and remove trees in residential properties had not come before the courts.

Dr. Graham is likely simply repeating the official Department's line at the time. Deputy Commissioner Craig Meyer testified in May 2001 in Miami-Dade District Court, that the Commissioner decided on the moratorium, in part, due to opposition by residents. It is likely that at the time of the moratorium, the Department was already considering a much larger radius, but was concerned about lawsuits. The Valera lawsuit was certified in May 1998 by Judge Amy Dean and lasted until it was dismissed by the Third District Court of Appeals in April 1999.

Dr. Graham's portrayal of the activist homeowner who could influence policy is at odds with what was reported in the media. At least by 1999, it appears homeowners had grown tired of complaining to the Department. The class action lawsuit had been dismissed in April 1999, and beyond letter writing, there was little means of protesting. Evidence of this comes from a town hall meeting organized on June 18, 1999 in Miami Springs, Florida, to answer questions related to the restart of the program following the moratorium. The Miami Herald reported that only 10 people attended the meeting.[6]

However, this poorly attended meeting was important in understanding the origins of the 1900-ft rule and fortunately covered by the Miami Herald. Present at the meeting were Dr. Timothy Schubert, Administrator of Plant Pathology from FDACS/DPI and Ken Bailey, the Director of Operations for the CCEP in South Florida. At the meeting, Dr. Schubert stated new research indicated a 1200-ft radius could capture 99% of the disease, while a 125-ft radius could capture only 30% of the disease. However, Ken Bailey indicated more frequent inspections might improve the effectiveness of the 125-ft policy.[6] The few residents attending the meeting probably left not knowing the direction of the Department— larger radius or more frequent inspections.

3. Opposition to the Program: Year 2000 forward

The opposition by residents to the CCEP intensified greatly with the 1900-ft rule. With the 260 acre circles, it is likely that many residents would have no idea where the offending infected tree or trees were located. All they could see is healthy trees in their area being cut down.

Most likely, the inspectors were treated much better by residents than the privately contracted cutting crews. If a tree was filled with fruit, the protocols called for the cutting crews to gather up all fruit and throw them into the chipper. More than a few times, the cutting crews seemed to be collecting fruits for themselves.

The 1900-ft policy also included removals of tree trunks, which meant bringing in heavy stump grinders to the backyards of residents. Damage occurred to fences, irrigation, landscape lighting systems and other landscape features. Often it was observed that stump grinding equipment would not be used, as it simply would not fit through the gates of residents or cause a hazardous situation. The limestone base of many backyards means small chips of coral rocks would fly when stump is ground. The most dangerous action occurred when stump grinding equipment severed a gas pipeline. Fortunately, the gas was not ignited, but it was a close call.

Many residents felt their rights had been violated as the cutting crews arrived with chain saws, but they really did not know how to protest the action. Letters or emails were sent to city commissioners, mayors, and representatives in the Florida House and Senate.

The list of what was wrong with the eradication program was long, however it generally included the following:

1. The 1900-ft rule was not based on science. No technical studies had ever been made public to justify the 1900-ft rule prior to its enactment. No studies were ever done to show that the trees within the eradication circles would get canker after the infected tree was removed. Further, no studies or reports were ever presented before year 2000 showing the prior eradication program with 125-ft circles was ineffective.

2. There were better methods to control citrus canker rather than destroying healthy trees. These methods included the use of copper based chemicals (fungicides) whose application had to be properly timed, windbreaks, growing more resistant cultivars, and removals when necessary of infected trees. Growers were already using fungicides to control other diseases. Canker only meant more frequent spraying.

3. This program was simply an over reaction of the Department to satisfy the citrus industry. Instead of eradicating the disease in the groves where it could do the most damage, the Department was destroying healthy citrus trees in peoples' yards where little or no

compensation was available. Also, eradication of residential trees benefited the citrus growers, as it reduced the supply of free backyard citrus fruits.

4. The program was in violation of basic rights of residents. The inspections of homeowners' yards and subsequent cutting of trees violated the Fourth Amendment Rights against unreasonable searches. The lack of fair compensation was a violation of Fifth Amendment. Finally, the lack of any viable legal recourse for owners was a violation of due process contained in the Fourteenth Amendment to the Constitution.

5. To gain support for the program, the Department had exaggerated the impact citrus canker has on the health of trees. The citrus canker disease does not kill the citrus tree. In fact, citrus fruit with canker lesions is less attractive, but the disease does not diminish the quality of the fruit. The issue was an economic one for grove owners, since the citrus from diseased trees could be used to make juice, for which they would receive less money than fresh fruit.

6. The 8 billion dollar "economic impact" value was a product of economists' imagination. The real "maximum" value is closer to 1 billion based on the "on tree" revenue value of all commercial citrus grown in Florida. The 8 billion value was based on an economic model, which included all possible revenues from associated businesses. However, the scenario that the citrus industry would be eliminated and new business would never fill the void is absurd. In fact, new housing has per acre a much greater economic impact than groves. Employment values of 60,000 were also the product of modeling, and actual full time employment likely closer to 10,000 to 20,000 employees.

7. The over reaction of the Department, in clear cutting large areas of South Florida, was motivated by trade war fears, particularly from China. The US had banned Mexican fruit imports for years based on Medfly concerns. Retaliatory embargos from other countries and the likely event that the USDA would restrict the export of fresh fruit to other states were the real motivating factors behind the CCEP.

8. The program was doomed to failure in the same way the two prior "canker wars" had failed. Canker was endemic to Florida. The time to discovery (two years or more) meant the inspectors would be unable to find many of the infected plants, because the trees would not yet show symptoms of canker. Any apparent success would be short lived as residents replanted their trees, and those remaining infected trees (remnant trees) would be the sources of the next generation of infected trees.

9. The citrus industry including the entire supply chain to the grocery stores benefitted from the residential eradication program as residents were deprived of their backyard supply of citrus. Also, the citrus grove owners who had their grapefruit trees removed, could replant with rootstock which would be resistant to citrus tristeza virus. The sour orange rootstock was susceptible to this virus.

There were numerous other arguments against the program. One resident, Marion Henderson, suggested the Department needed to use chemicals which improve a citrus trees' ability to naturally resist infection (systemic acquired resistance). Her arguments were bolstered by scientific papers on systemic acquired resistance of citrus plants. Other residents, including Helen Ackerman felt the Department should be testing new and safe chemical treatments, as suggested by the companies involved in canker research, including the bactericide Oxidate product from AgSolutions, Inc.

Broward and Miami-Dade County attorneys, Andrew Meyers and Robert Duvall, respectively, carefully reviewed the legality of the program. They found many legal issues with the program including basic constitutional violations including unreasonable search (Fourth Amendment), and lack of compensation for property destroyed (Fifth Amendment) and lack of recourse or due process (Fourteenth Amendment).

Broward County attorney Andrew Meyers uncovered more legal problems than realized by residents. The program had circumvented the normal "rule-making" procedures, which required public hearings to allow comments from residents. As mentioned in the introduction, the notice given to residents (Immediate Final Order) was faulty because it never indicated that a homeowners' trees were going to be cut down because they were within the 1900-ft circle. Thus, there seemed no legal way for the courts to block the cutting of trees. The legal issues were heard in the Broward County Court in October- November 2000 and Miami-Dade County Court in May 2001. The Broward and Miami-Dade cases are reviewed in Chapter 5.

Resident in some cases sought their own lawyers to sue the Department for what they considered to be a violation of their rights. Dr. Westphal sued the Department in year 2000, on the basis that the program violated the right of entry without a warrant. Others filed lawsuits, hoping to block the destruction of their trees.

Other residents protested in more conventional manner, through letters to the newspapers and in town meetings. Numerous letters to the editors were published. Both the Miami Herald and the Sun Sentinel (Broward County newspaper) came out strongly in favor of the program.

Miami-Dade County Commissioner Katy Sorenson sponsored a resolution supporting compensation to lime grove owners for their losses. The resolution was passed by the County Commission on March 9, 2000.

Not everyone protested in a conventional manner. One resident painted his house orange in protest. In year 2000, a parade was held at the end of the year, in Miami, Florida, called the King Mango Parade. This Parade was started in jest, so all those who were rejected from the very commercial Orange Bowl Parade could participate. To protest the eradication policy, a resident had put an orange tree on a hand cart, with a sign — Citrus Canker, keep 1900-ft away.

Letter from a Concerned Citizen

One opponent of the program, Dr. Melvyn Greenstein wrote a letter which appeared in Pinecrest Community Newspaper and was posted as part of the Pinecrest Tribune in 2003. Excerpts are provided below:

Which is more invasive, canker or the state?

> Continuing their rampage to rid Florida citrus of unsightly canker, agriculture officials are ready to scour private property for not only infected trees, but all healthy trees (the state calls these exposed trees) within 1,900-ft of an infected tree.
>
> The Florida Department of Agriculture wants you to believe citrus canker is a menace of utmost urgency and the best was to protect the big commercial groves in Central Florida is by conducting a chainsaw massacre before canker gets there. People fighting the state think that approach is like throwing the baby out with the bath water. The fact that their exposed trees keep thriving has only deepened their distress and mistrust.
>
> To other special interests, not even the tourism sector has the singular clout of big agriculture. No other industry could have persuaded government to deploy a taxpayer-funded army to wipe out millions of perfectly healthy orange, lime and grapefruit trees.
>
> The most frequently asserted argument defending the citrus canker program is to claim it is based on science. Agriculture Commissioner Bronson referred several times to the so called Gottwald Report trying to justify the protocol to eradicate all citrus trees in a radius of 1,900 feet from an infected tree, calling it proof. The fact is, Gottwald's experiment did not recommend eradication and never even suggested it.
>
> A single experiment, if not repeated and confirmed by other scientists, is not scientific proof. The goal of the experiment was not to consider eradication, but to search for the distance of canker infection. The scientists participating in this research, Dr. Gottwald, Dr. Sun and Dr. Graham, examined only how far canker bacteria travels before it falls to the ground.

Further on, Dr. Greenstein writes:

> If a sector of producers is not viable to the competitive on the export market, the government may subsidize them to a certain extent for the sake of the economy. But to do so at the expense of individual homeowners, trampling on their constitutional rights and causing them immense emotional damages is unforgivable because the entire program is useless.
>
> So, where are we now? The homeowners believe that strangers with chainsaws aren't supposed to hop over your fence and start whacking down foliage without explicit legal authority. That view clashes head-on with the state's infamous edict that all citrus trees located within a 1,900-ft radius of an infected specimen must be destroyed, no questions asked.

> To thwart local homeowner resistance, the legislature last year legalized blanket warrants for entire counties and allowed state-hired choppers to basically go anywhere and cut down anything with the dubious infection zone. The legislators' measure is audacious in its reach and for its disregard of the Constitution's protection against unreasonable searches and seizures."

The "legalized blanket warrants" as Dr. Greenstein discusses, were a creation of the Department and when proposed in the legislature, were ridiculed by many senators. The law stipulated that once an infected tree was discovered on a property, this gave the Department probably cause to search every property in the county. An infected tree could be probable cause for entry into a property located 50 miles or more away.

Ordinary Citizens — Extra-Ordinary Perseverance

Broward residents Jack and Patty Haire saw the ever expanding quarantine areas and began to question the merits of the program. Was it legal for the Department to enter private property without permission and take healthy trees without compensation? Was it really necessary to cut down so many healthy trees? They researched citrus canker, its supposed harmful effects on fruit and the citrus trees. Canker just did not seem as harmful as the Department was saying.

Patty organized everything in a two-inch binder and they went to the Broward County Attorney's Office in downtown Fort Lauderdale near lunch time. Had they arrived 5 minutes sooner or later, they would have missed attorney Andrew Meyers. As luck would have it, they were in front of the perfect attorney for these issues. Mr. Meyers immediately showed a piqued interest in the legal issues surrounding the Department's actions. Being trained in contract law, Mr. Meyers soon spotted a serious problem— nowhere on the IFO notice given to residents did it specifically state that the Department intended to cut down healthy trees on a particular property. It gave the homeowner nearly no recourse to stop the action of the Department.

It was the just the beginning. The program was illegal on numerous issues. Jack and Patty Haire would ultimately become the lead plaintiffs on the citrus canker case as the case went from the Broward Court, to the Administrative Court in Tallahassee and back again to Broward County and finally to the Florida Supreme Court. As Jack traveled, Patty would communicate with others of what was going on, the issues at hand and where the next court hearing would be held. They were instrumental in holding together a loose coalition of program opponents. In the end, their contact list contained over 1,000 names.

The 1900-ft eradication program went on for six years, and during that time, Jack crisscrossed the state to speak at countless city, county and homeowners' associations meetings. Several times he traveled to Tallahassee and met with state legislators to make them aware of the falsehoods they were being told. It was very time consuming. Jack and Patty would have preferred spending this time enjoying the outdoors. Often Patty would lament to me that she wished someone else could take over this role and leave her to things she was drawn to such as tending to her wildlife habitat, native plantings and butterfly gardens. They made this commitment because there were basic constitutional rights on the line and a near certain precedent setting of the Department deciding another plant/tree needed to be destroyed for the good of a particular industry. Neither regretted the sacrifices made during their six year odyssey. There was an unwavering conviction that the Department should not get away with what they were doing.

Opponents Learn More from Other Sources

By year 2000, eradications were underway on the east and west coasts of Florida. Although several attempts were made to organize a single group to protest the Department's actions, no cohesive group emerged. Instead, the scattered opponents of the program would let their local representatives, such as their state senators and congressional representatives, town councils, or city hall officials and others know of their opposition to the program.

The portrait of citrus canker as a highly contagious and deadly disease, was soon countered by other information, which could be found on the internet. Public information officers were often presenting misleading information about canker at town hall meetings. If residents' trees were in the exposure zone, all their fruit would be contaminated and should never be shared.

A website was created called, www.saveourtrees.com which provided information on the provide current media stories and allowed residents to comment on the program. Through regular updates provided by an email system, residents could sign up to receive timely information. Other websites followed in protest to the program. I also created a website to provide more information for residents. My website posted all the court decisions on citrus canker and the scientific papers on canker research. Further my website provided copies of three papers written by Dr. Jack Whiteside from 1985 to 1988, critical of the prior attempts at eradication.

Lack of Legal Recourse

When residents found an "Immediate Final Order" (IFO) on their door, they were advised of legal measures they could take to stop the cutting of their tree. Unfortunately, these measures were costly, complicated and generally ineffective. If the Department made a mistake in identifying a tree with citrus canker, it was likely the tree would be destroyed before the court would intervene. By that time, the homeowner would have nothing handful of sawdust for the courts to save.

US citizens have a right to due process as guaranteed through the Fourteenth Amendment of the constitution, which states: "No State shall make or enforce any law which shall abridge the privileges or immunities of citizens of the United States; nor shall any State deprive any person of life, liberty, or property, without due process of law; nor deny to any person within its jurisdiction the equal protection of the laws."

Residents soon found there was little chance of successfully challenging the Department in court unless they could somehow prove their trees were not in the 1900-ft circle or that their trees were in fact, not infected with citrus canker. Legal challenges for compensation beyond what was offered by the Department was next to impossible as the Department could not legally pay any compensation without authorization by the legislature.

4. Opposition from Grove Owners

One of the erroneous claims of the Department, was that the citrus industry fully supported the 1900-ft rule. This may have been true of the large grove owners in Central Florida and the largest citrus industry organization, Florida Citrus Mutual. However, lime grove owners in the Homestead, Florida, located south of Miami, were generally not supportive of the 1900-ft policy and the manner in which it was implemented.

This division among grove owners is understandable. At the onset to the 1900-ft policy, the Department's goal was to stop canker in South Florida, so it would not spread to the large commercial groves located further north. There was considerable discussion of concentrating eradication efforts in Broward County, and even creating a firewall to limit further spread to the north.

Mr. Steven Sapp of Sapp Farms in Homestead took legal action against the Department, seeking an injunctive relief from the destruction of his healthy trees. The District Court denied injunctive relief, The Third District Court of Appeals ruled the Immediate Final Order, was not-defective on March 1, 2000.

Mr. Craig Wheeling, President of Brooks Tropicals, LLC in Homestead, Florida had serious doubts about many aspects of the program as he expressed in his letter of March 21, 2000 to Congressman Peter Deutsch, with copies to Commissioner Crawford and Department of Agriculture Secretary Dan Glickman.

It is clear from his letter, that there was confusion and poor coordination of efforts. Craig Wheeling was a member of the Technical Advisory Committee, so he was better informed than others. While the Department was claiming they were aggressively cutting down trees at the meetings, he was aware the Department was not cutting positive trees in his grove. In addition to the actions of the CCEP, strong doubts emerged of the scientific basis for the 1900-ft rule, primarily due to the lack of a documented study. Brooks Tropicals, LLC participated in the second trial for injunctive relief in year 2002, represented by Attorney Malcolm Misuraca.

USDA Provides an Opportunity for Growers' Comments

A series of comments were sent to USDA/APHIS from December 15, 2000 and early January 2001, from growers and others citrus organizations on the proposed program by USDA to pay grove owners for their production loss (Docket 00-037-2, Recovery for Loss Production, Published Dec 7, 2000). Brooks Tropicals argued that the payments should be higher. They also repeat many of the problems with the CCEP, in general agreement with the verdict of Judge L. Fleet in Broward. Other letters were sent to USDA, in agreement with Brooks' financial analyses.

In contrast, Dr. Jim Griffiths, Managing Director of Citrus Growers Association, to the USDA/APHIS, who in January 4, 2001, expressed the concern that the USDA compensation program might in fact be too generous, as he wrote,

> [USDA/APHIS] Proposal 00-037-2 actually provides higher than average market values, offers no opportunity for superior groves to be offered greater compensation because of their superiority, and in fact, the payments suggested are so high that they could actually encourage the unscrupulous operator to artificially spread citrus canker.

Dr. Griffiths letter is posted to the website.

The " Griffiths' tipping point" is defined as when the grove is worth more, with canker than without it. The unscrupulous operator would somehow acquire a tree with citrus canker or move a citrus canker infected tree to a location where it would the trees were not as productive.

Year 2000 would mark the beginning of an 7 to 8 year housing boom for Florida and the rest of the country with subprime loans. As land prices soared near urban areas, such as Homestead, many groves were likely worth more "dead" or plowed under or burned to the ground, than alive, with the generous payments from the USDA.

Grove Owners Letter to President Clinton

After sending letters to Congressman Deutsch, Brooks Tropicals had cc'd Secretary Glickman, and Governor Bush, and not getting much response, there was only one more level up to go - President Clinton. It seemed pretty extreme, but with one letter, Brooks could copy almost everyone. So, on July 18, 2000, a letter containing numerous grievances was sent to President Clinton, with copies to Governor Bush, Commissioner Crawford, Secretary of Agriculture Glickman, Miami Mayor Alex Pinellas and others. It is signed by Craig Wheeling and Neal Brooks of Brooks Tropicals, Herbie Yamamura of Limeco, Inc., Mark Philcox of Grove Services, Alcides Acosta of Acosta Farms and Steve Sapp of Sapp Farms.

The letter begins with a list of pests discovered in Florida after Hurricane Andrew in 1992. The implication is the USDA/APHIS is not performing well at interdicting pests. The letter further accuses FDACS of letting positive trees remain uncut in the groves for more than two months and delays in compensation. It opposes the quarantine of limes, and considers the field study, resulting in the 1900-ft insufficiently documented to judge its validity. The letter states, "The 1900-ft radius destruction zone is arbitrary and capricious."

Interestingly, in discussing inadequacies of the field study, the letter implies dissemination in residential areas "by dogs, cats, insects and their damage, meter readers, garbage collectors (and their trucks), lawn maintenance workers (and their workers), etc" which were not taken into consideration. The Department responds that in fact, these potential factors were taken into account.

The copy of the Department's response, is nearly illegible, as it was included within the attachments of Dr. Westphal's lawsuit. A transcribed copy is posted on the website along with a copy of the original.

5. The Department Responds to Program Opponents

The FDACS vigorously defended the program, both from an operational perspective and overall concept of healthy tree cutting. The following are selected excerpts from the Department website, and an article written by Dr. Schubert, from the FDACS/DPI and co-authored by other scientists involved in the CCEP. [7]

1. Citrus canker is a devastating disease that may be fatal to the tree.

At the November 14, 2001 Public Hearing, a resident asked if it is true that citrus canker kills a tree. The response of the Department is given below:

> Citrus canker has been recorded to kill young and mature trees under conditions optimal to colonization, reproduction, and spread of the bacterium and leading to high levels of disease incidence (e.g., Maldives Islands). In Florida, environmental conditions and host cultivars are present such that tree death may result under limited circumstances. More times, it is fair to say, that citrus canker (which flourishes in Florida) will cause susceptible citrus varieties to exhibit unacceptable levels of premature fruit drop; reduced tree vigor, and risk of infestation by secondary plant pests. Like with most biological questions, absolute black and white answers are impossible without leaving a person only partially informed. It is true that citrus canker can kill a citrus plant. The literature documents such an event [Plant Disease 73: 363-367 (1989)]. Canker can also kill young trees in the nursery environment. But it is not safe to extrapolate from that statement to say that "citrus canker kills trees", perhaps implying that death is always the result of infection, because that is not the normal situation. Nor is it accurate to say that citrus canker does not kill trees, because it certainly can.

2. Delays by court action allow canker to spread

In regard to the Broward court injunction, the Department's public relations officers responded quickly in November 2000 to the Broward court injunction, stating that they were within five weeks of completing a five year eradication program. In December 2000, there was considerable optimism in the Department, that the injunction would be quickly overturned by the District Court of Appeals, and full eradication program would begin again in February 2001.

3. The 1900-ft rule was based on sound science

The following is from a letter from Commissioner Bob Crawford to President Clinton on July 24, 2000. The Commissioner was responding to allegations made in a letter from South Florida grove owners on July 18, 2000.

> Concerning the validity of Dr. Gottwald's et al., citrus canker epidemiology research, work that is in prepublication form, it is important to point out that it has been reviewed by several of the world's leading plant disease epidemiologists. Dr. Gareth Hughes, of the University of Edinburg, Scotland, and a leading expert, has working closely with Dr. Gottwald in the analyses of his data and if anything, feels it may be conservative in its conclusion. As the research work has passed the initial scientific peer review, it is very appropriate to put it to good use. Regarding Dr. Gottwald's reluctance to share prepublication copies of this work, it is common practice in the scientific community as authors prefer to distribute final copies rather than draft copies of their findings for general distribution.

4. Unless eradicated, citrus canker would be economically devastating to Florida

As stated in article by Schubert et al.:

> After careful study, what initially may seemed to be a rush to judgement in favor of eradication is actually a well considered plan that takes into account the need to act swiftly to contain a spreading pest. This concern does not override the need to determine cost benefit ratios, for the planned actions. ... The decision to attempt eradication has considered cost-benefit calculations that clearly indicate this to be the wise choice in long run if the disease is caught early and action is taken quickly.[7]

The UF/IFAS did a cost benefit analysis, but this was done after the 1900-ft rule was implemented. The analysis is available online. It is noted that the minutes of the Task Force meetings do not show any discussion of a cost-benefit analysis to justify the 1900-ft rule.

5. Experts conclude that unless citrus canker becomes endemic to Florida, eradication is far better than control methods.

This argument was used in both Canker War II and III. Since there was a nearly 50 year period between the first and second Canker Wars, the claim was that eradication has been successfully done once and it can be done again. The Department argued in Canker War II, that Argentina had failed in eradication, so there only choice was control, which would continually be bad for growers.

6. Concluding Remarks

Residents of South Florida had no voice in policy making. The task force included Department and USDA officials, citrus industry association leaders and UF/IFAS scientists. Residents had nearly no means to stop the destruction of their trees through the appeals process in the courts.

Prior to January 2000, and implementation of the 1900-ft rule, it is very likely the Department fully recognized there would be backlash from residents. However, they had a plan. The public relations effort included excellent contacts with the media and civic leaders. There would also be support from the State Agricultural Extension offices. Florida Citrus Mutual was onboard with the program, particularly with the compensation plan. The final obstacle was legal, and with the dismissal of the Sapp case on March 1, 2000 by the Third District Court of Appeals, the road, with legal, financial and political impediments, finally seemed open.

The public relations officers of the Department would state unequivocally that not a single plant pathologist in Florida disagrees with the 1900-ft rule. The truth of the matter would later come out in court testimony. Retired plant pathologists did disagree with the procedures in the field study. Other plant pathologists working at state supported universities preferred not to voice their doubts. It was not their program and it was easy to look the other way.

What the Department did not fully anticipate, was that the county attorneys from the legal offices in Miami-Dade and Broward counties would file lawsuits on behalf of their residents. Despite all the failed challenges in the past, these new lawsuits would gain traction, and not go away easily. The few months of delay in implementing the 1900-ft rule, had cost the Department dearly.

The residents who attended meetings various city hall meetings, were likely to encounter only Public Relations officers, such as Ms. Liz Compton, Richard Miranda or Richard Fagan. These officers would immediately profess to not knowing any details of the epidemiology study. They would provide brief well rehearsed summaries of how a team of scientists from well known institutions, including the University of Florida, University of Edinburgh and the USDA, determined that 95% of the bacteria falls to the ground within 1900-ft of an infected tree through tracking disease movements in controlled test sites for over a year.

The Department would rarely admit to alternatives, such as control measures, as had been argued by Dr. Whiteside during Canker War II. The arguments between control and eradication, should now be considered academic as far as Florida's situation is concern, since the USDA concluded in 2006 that eradication is infeasible.

But something felt very odd in year 2000. There was not a single document on the research leading to the 1900-ft rule. The public information officers were telling everyone, that the study was in the hands of Dr. Gottwald, and that he had not yet completed his report. Dr. Gottwald and the other researchers involved in the study never made any presentations to Florida residents.

The Department likely did not anticipate the response of the South Florida lime grove owners. Unlike residents, they had the resources to file lawsuits. They decided that since they were being asked for comments on USDA regulations, it opened the door to address their grievances to the US Secretary of Agriculture. Not to be outdone, the Commissioner of Agriculture responded with a letter to the President Clinton.

The grove owners' letters were obtain through a Freedom of Information Act request. The most surprising one, was from Dr. Griffiths, representing the smaller citrus growers, who stated in general, that the compensation to grove owners was in fact too generous, and unscrupulous grove owners might take action to spread canker to their groves. His 40 years experience shows how anyone, given the opportunity would take advantage of government handouts.

The protest by the lime grove owners in the Homestead area seemed to diminish quite quickly, perhaps as they began to receive their checks for lost production, and so they could recover from the eradication program. The approximately 3000 acres of lime groves are gone today, and at least half of this loss was due to the eradication program. Much of this land likely was sold to developers.

Protesting the program was very difficult. It was difficult to know who to contact and lodge complaints against the program. Also, the damage done per resident was not large, but because 850,000 citrus trees were destroyed, in the aggregate it represented hundreds of millions of dollars of destroyed property. Just the fruit from one large orange tree, can be valued at several hundred dollars. Many residents participated in the Public Hearings, occurring in October to November 2001 and in town hall meetings held in years 1999 and 2000. Others had letters to the editor published in local papers. Many complained directly to FDACS.

Short Notes on the Website

SN 1.2 Dr. Whiteside's Contribution to Citrus Canker Research
SN 1.6 What is an Exposed Tree?
SN 1.8 Overwhelming Emotions, by Patricia Haire
SN 1.9 Citrus Canker and the Giant Yellow Swallowtail, by Roger Hammer
SN 3.1 The Case of the Dirty Scalpel
SN 4.1 An Essay on the Citrus Canker Eradication Program by Dr. Peter Haransy

5. LEGAL CHALLENGES

> The Department has spent taxpayers' funds to retain the finest of legal talent to support its cause and has invested nothing of significance to allow the people it is supposed to represent an opportunity to litigate on an even judicial or administrative playing field.

Broward County Court, Judge Fleet in his decision, November 17, 2000

> Does a single citizen really have the "right" to harbor on his property, potentially a disease that could wipe out every single citrus tree in Florida?... My amateur's reading of past court cases is that no person has the "right" to harbor a public threat.

Howard Troxler, St. Petersburg Times columnist, June 5, 2002

> "Every time you pull another layer off of the eradication program, it gets worse," he said. "The Department of Agriculture is an equal opportunity violator. It violates state statutes, court rulings and private property rights, seemingly without concern."

Andrew Meyers, Broward County attorney, Sun Sentinel, June 28, 2003

---

1. The Finest Legal Talent Created an Illegal Program

On November 17, 2000, Judge Leonard Fleet of Broward County ruled the Florida Department of Agriculture and Consumer Services ("the Department"), had, in no uncertain terms, violated Florida laws in the eradication of healthy citrus trees and ordered healthy tree cutting in Broward County be stopped immediately.

Residents felt elated as they exited the courtroom. Surely, they had put a stop to the program. However, one of the attorneys quietly told them, "This is the just the beginning." The Broward case would last four years, ending in the Florida Supreme Court. Private class action lawsuits for additional compensation are still in the courts as of 2017.

On February 3, 2000, a confident Deputy Commissioner Craig Meyer told the Citrus Canker Technical Advisory Task Force, that anyone who attempted to challenge the program would be stopped initially through a bond hearing as follows:

> We will insist on a bond and we insist that the value of the bond is somewhat related to the value of the citrus industry that is so high, no one will be able to leap that hurdle. That is our first line of defense.

The University of Florida/IFAS had estimated the economic impact of the citrus industry at 9 billion dollars. The multi-million dollar bond request never happened. Judge Fleet did not require the litigants to post bond in the legal challenges.

The Department portrayed those against the program, as a few residents who were more concerned with their backyard trees than the millions of trees in commercial groves. In fact, the legal challenge involved attorneys from Miami-Dade and Broward counties and 8 municipalities.

Old timers in Florida will tell you there are four phases in citrus canker eradication: (1) Lofty goals of complete eradication, gross underestimation in terms of money and time required, coupled with exaggerated claims of the destructiveness of the disease, (2) A reality phase which consists of some progress in some localities, followed by lapses in funding and then even more discoveries from new intense surveys, (3) Renewed optimism following an infusion of new money and (4) Lawsuits that can go on longer than the program itself.

The eradication program begun in 1995 met or exceeded many of the old timers' expectations particularly in terms of lawsuits. The lawsuits from the 10 year eradication program lasted more than a decade. However, this time around there were two sets of lawsuits. One set was from the municipalities, namely Broward and Miami-Dade counties, suing on behalf of their residents, challenging the legality of the program, with the aim of forcing the Department to work within the Florida Statutes.

Another set of lawsuits came from private attorneys challenging the lack of full compensation (inverse condemnation) for the destruction of healthy trees and demanded additional compensation through class action lawsuits. This set of lawsuits will likely continue into beyond 2017.

Unlike prior eradication programs, lawsuits came soon after the first healthy trees were cut down. A lawsuit initiated in 1997 in Miami-Dade County (the "Varela case"), and resulted in the certification of class action of affected Miami residents for compensation (called "inverse condemnation") filed by Miami attorney John Ruiz. This case (*FDACS* v. *Varela*) was dismissed in 1999 by the Third District Court of Appeals based on a prior case ruling (the "Polk case"). Previously, the Appellate Court had ruled that the healthy trees in a nursery within a 125 ft eradication circle were "worthless" because they had no market value. The nursery trees were worthless as they were considered exposed, subject to a quarantine and could not be sold.

To many program opponents, the primary issue was one of state's privileges verses citizens' rights. Could the state just come in and cut down healthy trees? Residents were told that this was a joint project between the USDA and the State of Florida. Did anyone seriously want to take on both Florida and Washington to protect a couple of citrus trees? Amazingly, they did with the support of their local government's legal office. These lawsuits were asking very good questions and looking at the justification of the program. Where was the support for the 1900-ft policy change? Was it really data-driven or just for the sake of convenience?

Without these cases, the information on the 1998 Florida field study would have been very limited. There might have been many more documents on the Florida field study through the formal discovery process, if the Florida Supreme Court had not effectively ended all challenges in 2004 to the 1900-ft policy. The level of examination of the 1900-ft rule was limited to bearing "a rational relation to the legitimate legislative purpose of safeguarding ... public welfare" and not to the validity as a result of scientific inquiry.

It is suggested that the early unsuccessful legal challenges to the program (the Varela case from 1997 to 1999) delayed the implementation of the 1900-ft rule, because the USDA was reluctant to participate in a greatly expanded program with an ongoing class action lawsuit in Miami-Dade County demanding compensation.

The scope of this chapter is limited to a brief historical account of the legal issues and the manner the court rulings affected the eradication program. There is no attempt to provide a legal commentary on these rulings in context of broader issues of interpretation of statutes and case law. The Appellate and Supreme Court rulings likely will impact future cases, in the areas of search and seizure, fair compensation and the need for due process, but this is beyond the scope of this chapter.

The rulings will be maintained under the online supporting documents website, along with links to select articles which discuss the broader significance of the cases.

2. Short Summaries of Legal Issues

Legal challenges to the program were not always greeted sympathetically. As stated in the opening quotes, Mr. Troxler accepted the argument that the harm caused by citrus canker would be so severe that leaving one "exposed" tree standing might cause the demise of the citrus industry. The Department claimed these exposed trees would eventually show symptoms of canker infections and become the new sources of bacteria. Thus, the challenge of the policy was deeply rooted in a couple of simple questions: How certain can one be that the healthy tree in the eradication circles would become infected? Where was the evidence?

The legal problems of the CCEP ranged from minor problems with inadequate notifications to larger issues of entry into properties without warrants or probable cause, inadequate compensation and lack of scientific justification of the 1900-ft rule.

In year 2000, an obvious flaw was the "Immediate Final Order" (IFO) intended to be a warning notice to homeowners of tree cutting on their property, which failed to provide any specific notice was given that a resident's trees were going to be eradicated. Instead there was a general statement of policy– all citrus trees in the 1900-ft circles would be cut down.

The Broward Court barred the cutting of healthy trees in November 2000 by issuing an injunction. This did not stop the Department from cutting infected trees in Broward nor stop healthy and infected tree cutting in Miami-Dade County.

The Department's decision of not correcting what seemed minor flaws made little sense to everyone but seasoned attorneys. It is suspected the lack of information in the IFO's meant the residents could not appeal any action of the Department as they could not show pending harm to their trees.

Other issues that came before the courts were:

- Search of residents' yards without permission or search warrants (Fourth Amendment rights)
- Effect on the environment in destroying over a million citrus trees
- Validity of the studies supporting the 1900 ft rule
- Right to full and fair compensation (Fifth Amendment Rights)
- Right to "Due Process" of residents, to legally challenge the destruction of property in court (Fourteenth Amendment)

In general, the Department's right to eradicate diseased trees was acknowledged and not an issue. The prior eradication program only entered commercial establishments, so the entry without permission was a new legal issue to the program.

Appellate Judge Jorgenson, in the case of David Markus, et al., v. FDACS (3D00-2762) made it abundantly clear how the residents had virtually no means to challenge the action of the Department, in the District Court of Appeal, as follow:

> Property owners as well as judicial tribunals are struggling with the issue of how and why the Department of Agriculture embarked on its dogged obliteration of the healthy trees back (or front) yard citrus trees. The frustration of challenging this policy, either in a Chapter 120 proceeding or before this court are staggering. Both infected and condemned trees are removed and ground to dust before any meaningful action can be taken by the property owner. The "final agency order" is nothing but a "Dear Resident" form from the Department of Agriculture. A "record on appeal" is an oxymoron. There is no record. Hence there is no meaningful appeal. We find that situation unacceptable as a matter of law, policy and principle, yet must affirm.

In the prior eradication program, only citrus within groves and nurseries were eradicated. The Department had the right to enter commercial properties for the purposes of inspections and eradications of infected and exposed plants.

The requirements for search warrants for residential properties was first addressed in a lawsuit filed by Dr. Westphal in July 2000, a retired philosophy professor at the University of Miami. In 2001, the City of North Miami and Miami-Dade County file a lawsuit on the basis of illegal searches. Although trial was decided in favor of the municipalities, it was later overturned based on lack of standing (the Plaintiffs could not demonstrate actual harm as a result of the residents' trees being cut down.

The rulings in the Broward and Miami-Dade Cases are provided in electronic format (pdf files) in the supporting documents website.

3. Federal and State Constitutional Rights

Fundamental to the challenges brought to the CCEP where basic constitutional rights. The US Constitution and the Bill of Rights may be cited in support of any case brought to court. The applicable rights guaranteed to the people under the Bill of Rights and Florida Constitution are:

Bill of Rights, Fourth Amendment

The right of the people to be secure in their persons, houses, papers, and effects, against unreasonable searches and seizures, shall not be violated, and no Warrants shall issue, but upon probable cause, supported by Oath or affirmation, and particularly describing the place to be searched, and the persons or things to be seized.

Bill of Rights, Fifth Amendment:

No person shall be deprived of life, liberty, or property, without due process of law; nor shall private property be taken for public use, without just compensation.

Fourteenth Amendment (Second Sentence)

No State shall make or enforce any law which shall abridge the privileges or immunities of citizens of the United States; nor shall any State deprive any person of life, liberty, or property, without due process of law; nor deny to any person within its jurisdiction the equal protection of the laws.

Florida State Constitution, Article 1, Section 12

The right of the people to be secure in their persons, houses, papers and effects against unreasonable searches and seizures, and against the unreasonable interception of private communications by any means, shall not be violated. No warrant shall be issued except upon probable cause, supported by affidavit, particularly describing the place or places to be searched, the person or persons, thing or things to be seized, the communication to be intercepted, and the nature of evidence to be obtained. This right shall be construed in conformity with the 4th Amendment to the United States Constitution, as interpreted by the United States Supreme Court. Articles or information obtained in violation of this right shall not be admissible in evidence if such articles or information would be inadmissible under decisions of the United States Supreme Court construing the 4th Amendment to the United States Constitution.

Florida State Constitution, Article X(6):

(a) No private property shall be taken except for a public purpose and with full compensation therefore paid to each owner or secured by deposit in the registry of the court and available to the owner.

4. Key Results from the Legal Challenges

The Florida Supreme Court decision of February 14, 2004 ended all legal challenges to the 1900-ft policy. It was neither a total win nor loss for either side. But the court's decision gave the green light to continue cutting both infected and healthy trees.

Key results for residents were:

- Search warrants for entry into yards were required if owners did not give their permission to inspections and eradication. County wide warrants were ruled unconstitutional. Warrants with multiple addresses and electronic signatures were permissible. In the lower courts, the Department had failed to demonstrate that "exigent" circumstances existed for the canker program. This was decided in part, based on evidence produced at the Broward Court trial that infected trees were left standing for long periods after discovery.

- Removal of residential trees in a 1900 ft zone was considered "a taking" by the Florida Supreme Court, providing a basis for a class action lawsuit for compensation. The ruling did not completely settle the compensation issue. It did not state what was destroyed, a worthless tree because symptoms of canker would eventually appear (Department's viewpoint) or a healthy tree (opponents' viewpoint).

Key results for the Department were:

- The Department has the power to destroy both healthy and infected trees in controlling an epidemic. Thus the courts upheld the Department's general "police powers." The Courts did not accept the argument that the average homeowner had little recourse to contest the destruction of his property or the value of the destroyed trees. The Department was ordered to provide the owner with relevant information on the IFO.

- The level of examination of the 1900-ft rule was limited to bearing "a rational relation to the legitimate legislative purpose of safeguarding ... public welfare."

The last key result, was a devastating blow to any continuation of the Broward Court trial. Attorneys for the municipalities had argued that they should be able to use a much higher standard of strict scrutiny, where it would be the burden of the Department to demonstrate that eradication of healthy trees was within the narrow limits of necessity. With compensation, the Florida Supreme Court ruled that the "rational relationship" standard applied.

Scientific Judgements on the 1900-ft Policy

The courts did not validate the 1900-ft policy. The Appellate Court reversed the lower court ruling, stating the 1900-ft policy was not arbitrary or capricious, based in part by the peer reviewed publications on the Florida field study. A thorough critical review of the Florida field study never occurred in court.

Discovery including depositions of the scientists was getting underway in December 2001 for a hearing in Administrative Court, but was cut short when the legislature passed the new citrus canker bills, signed into law by Governor Jeb Bush.

After healthy tree cutting was shut down for a second time in May 24, 2002 with a temporary injunction, the Department could have gone ahead with a trial to demonstrate the 1900-ft rule satisfied the higher standard of strict scrutiny and they were destroying only those trees within the narrow limits of necessity. If they had taken this option, and prevailed, compensation might not have been required.

5. Broward County Case 1

Case #: 00-18394(08), Decided on Nov 17, 2000

Legally, the municipalities can not sue in court unless they can prove pending harm as a result of the Department's action. The lead plaintiff in these Broward Cases, were residents who lived in the county and owned citrus trees within the eradication circles.

The reference to Broward cases 1, 2 and 3 is done for convenience. These cases have the same case number for filing purposes. The lead attorney in the Case 1 was Mr. Andrew Meyers, an attorney with the Broward County's legal office. In the second case, lead attorneys included Mr. Robert Duvall III from the Miami-Dade County legal office.

Issues involved: The 1900 ft policy was never enacted as a rule of the Department, an abuse of Departmental authority, IFO did not specifically state that owners trees would be cut down and no practical means to contest the cutting in court violation of Florida and US Constitutional rights against unreasonable searches, there was no scientific or technical basis to support the Department's contention that 1900-ft circles would eradicate canker and there was no support to show the goal of eradication would be attainable.

Final Outcome: The complaint on the inadequacy of the IFO, ultimately was upheld by the Administrative Court. Appeals by the Department resulted in nine months stoppage of healthy tree cutting. The Department was forced to go through rule making. The initial rule they adopted was ruled by Administrative Court to be a violation of Florida Statute 120 as it exceeded the authority provided under Florida statutes.

As both FDACS' rules and Florida statutes were changed in 2001, a number of issues (search warrant issues, validity of the science) returned to the Broward County courthouse, beginning Broward Case #2. In support of the program, the legislature adopted a new form of warrant, the "agricultural warrant" that would cover all residential properties in a county and give the Department of Agriculture (DOA) broad police powers.

Details: The legal journey that began with initial filings in October 2000, challenging the legality of the program. An evidentiary trial was held in the Broward County court house before Judge Fleet for three weeks.

Judge Fleet ruled on November 17, 2000:

- The IFO's were invalid because it did not tell owners that their trees were being cut.
- The 1900-ft rule was unenforceable because it had never been enacted as a rule or law of Florida.

Judge Fleet harshly criticized the Department, for excluding residents in the policy change, and changing their rules so the resident would get the absolute minimum information on the IFO. He indicated how difficult it was for homeowners to contest the destruction of their trees:

> Plaintiffs presented Ms. Caroline Seligman as a typical homeowner who sought to utilize the judiciary to stop what she perceived to be the wrongful destruction of her citrus trees. Ms. Seligman contacted the local office of the Department, but received little or no help. To her great fortune, Ms. Seligman met an unidentified attorney whom she said was Board Certified in appellate law. The unnamed attorney helped her draft a petition for stay which she later filed with the District Court of Appeals in West Palm Beach. The Department's response to the clerk issued stay order was a nine page brief with six exhibits attached. As required by law, Ms. Seligman would be required to respond to the Department's appellate brief were she to have any hope of preventing the destruction of her citrus trees. If the Department deems it necessary to retain the services of such distinguished law firms as Adorno & Zeder and Greenberg Trauig, how can one not trained in law adequately defend one's position in this complex case? The question clearly answers itself.
>
> The disparity in representation between the Department and a home owner is obvious and drastically tilted in favor of the Department. The Department has spent taxpayers' fund to retain the finest of legal talent to support its cause and has invested nothing of significance to allow the people it is supposed to represent an opportunity to litigate on an even judicial or administrative playing field.
>
> The cavalier attitude of the Department towards the rights of the general populace is not acceptable to this Court and should not be acceptable to any other reasonable judicial or administrative body.

Further in his opinion, Judge Fleet ruled:

> President Truman was right when he said secrecy and a free democratic government don't mix.

On the issue of whether the Department's 1900-ft rule is "good science", Judge Fleet states:

> In determining whether the Department's destruction policy is supported by competent, substantial evidence, this Court may not reweigh the evidence or substitute its judgment for the judgment of the Department. Nor is the Court entitled to second guess the Department's decision of engage in a battle of the experts.

Further, in summing up testimony on the science issue given at trial, Judge Fleet states:

> While Dr. Bailey, Dr. Whiteside and Dr. Wutscher certainly raise valid concerns about the basic assumptions underlying the field study, the legitimacy of the 1900 foot rule, the chipping method, the 95% capture rate and whether citrus canker can actually be eradicated in Florida, this Court does not have the legal authority to conclude the Department did not rely upon substantial, competent evidence upon which to base its present course of action.

The Department claimed Judge Fleet had shut down the entire program, however this was not true. They had the right to continue to inspect properties, and cut infected trees in Broward County and healthy trees cutting could continue in Miami-Dade County. There was no ruling on the search warrant issue because of the other problems in the program had effectively stopped healthy tree cutting in Broward County.

Instead of limiting the cutting, the Department chose to shut down the entire program at this point in both Miami-Dade and Broward for reasons that have never been made entirely clear. It is suspected that at this point, the Department feared a similar trial could begin in Miami-Dade County if healthy trees were cut. It is also suspected that the Department was running out of money.

Fourth District Court of Appeals (June 20, 2001)

Department subsequently appealed the Broward decision. The Fourth District Court of Appeals ruled on June 20, 2001 that all administrative remedies had not been exhausted, so any challenge would have to be heard in Administrative Court. Their decision lifted the injunction against healthy tree cutting. At that point, it appeared to many that cutting of healthy trees was going to start again. However, cutting did not resume.

The Department's appeal demonstrated how incredibly complex our legal system can be sometimes, as one would think the proper place for legal challenges to healthy tree cutting should be in the local court houses. Instead, the 4th DCA told attorneys in essence, this case had to be decided in Tallahassee, Florida, approximately 500 miles away.

Administrative Court (July 31, 2001), First District Court of Appeals (August 17, 2001)

An expedited hearing was held in July 2001 in Administrative Court in Tallahassee. The hearing lasted one day. In a very detailed opinion, handed down on July 31, 2001, Administrative Judge Laningham reviewed the history of the program, and various laws and rules that had been passed over the last 20 years to authorize the Department to eradicated citrus canker. Following this review, and the evidence presented in court, Judge Laningham declared:

> The Department's existing definition of "exposed trees" constituted an invalid exercise of delegated legislative authority. The law allowed cutting of healthy trees that harbored bacteria and in time would develop canker symptoms, while the rule allowed cutting of those trees that in time would likely to develop symptoms.

A quick-fix proposed "new rule" which had its first public hearing the previous week, was found to be in violation of Florida Statute 120.54. The new rule allowed the Department to make exceptions to the 1900-ft policy based on vague risk assessment procedure. The Judge did not rule against risk assessment, only in the vague and in-house manner which allowed for judgments at the whim of the Department.

It was obvious that the Administrative Court kept a tight reign on the Department's authority per Florida Statutes. Ironically, in its appeal of the Broward decision, the Department had argued vigorously that this was a matter for the Administrative Court to decide. It is likely the Department wanted the case taken away from Administrative Court and Judge Laningham, even if it meant months of delay in the program and a return to Judge Fleet's court room.

On August 17, 2001, the First District Court of Appeals stopped healthy tree cutting based on the outcome of Administrative Court, the eradication program would again be enjoined against cutting healthy trees, awaiting the completion of rule-making procedures.

New Rule for Department (September to November 2001)

It seemed initially, the Department desired to put the hearings on a fast track, by conducting a single hearing in Tallahassee. A short time later, they scheduled a series of public hearings in Broward and Miami-Dade with the last public hearing in November 2001.

It is possible that the numerous public hearings were simply a delaying action, because the overall plan was to change both the Department's rule and Florida's law, and the Florida legislature was not yet in session. The Department had hoped to change the laws regarding citrus canker and requirement of warrants, to stop all legal actions.

Since Judge Laningham had declared the "new rule" a violation of Florida State Law 120, public hearings were held on the "new-new rule." This rule did also showed some legal ingenuity. The exposed tree was once again redefined. It did not have to harbor bacteria nor be likely to do so. Now, an exposed tree was defined geographically; it was any citrus tree within the 1900-ft eradication circle.

The new rule was adopted in November 2001 and immediately challenged by Broward County and four municipalities contending there were many violations of law including: the right to enter, inspect and remove trees without owner's permission or search warrants, insurmountable problems in challenging an IFO, and the 1900-ft rule was unsupported by science.

A trial in administrative court was in the preliminary stages in December 2001. Judge Laningham would have presided over this case. The Department was not pleased that the Judge had meticulous parsed through each phrase in the rules of the Department before ruling that the Department had exceeded their authority in a 73 page opinion. The Department successfully had Judge Laningham removed from the case based on very scant evidence of a single letter that implied he had spoken to someone about the case while it was active.

New Florida Laws Passed

In December 2001, as discovery proceedings were underway for the case in Administrative Court, the legislature began debate on a bill designed to embedded the 1900 ft rule in the law. This action was at the request of FDACS designed to circumvent the legal challenges against the Department in Administrative Court.

The new citrus canker bill redefined an exposed tree as any citrus tree within 1900 ft of one infected with citrus canker and authorized eradication of all trees within the circle. Also, the bill included an amendment to FS 933 allowing for "agricultural warrants" which would allow for entry into any properties that were within a counties where canker had been discovered.

On March 7, 2002, the bill was introduced into the Senate and a spirited debate ensued. Senators Posey and Pruitt spoke in favor of the bill emphasizing the need to ahead of the spread of canker. Senators Geller and Wasserman-Schultz introduced an amendment restoring the bill to its original form which required a hearing prior to any warrant being issued. Senator Geller stated that the bill had been completely re-written within the Finance and Banking Subcommittee, and asked Senators to restore the bill as it had been passed the Agriculture Subcommittee. A second amendment was introduced by Senator Villalobos, to eliminate blanket warrants. Senator Villalobos spoke for the need to respect fourth amendment rights against unreasonable searches. Both amendments failed to pass, and Governor Jeb Bush signed the bill into law on March 18, 2002.

Also, as part of the new citrus canker laws, the legislature also passed Florida Statute 581.1845 which included compensation for residents. It is suspected the Department understood that with compensation, the *Corneal* decision would apply and this could end the judicial review of the validity of the 1900-ft rule by application of the "strict scrutiny standard." In essence, they were buying their way out of a court case under which the eradication policy would be examined under the narrowest limits of necessity.

There was immediate negative reaction to this law. Comments from lead attorney for Broward, Andrew Meyers have been posted on the online supporting documents website.

6. Broward County Case 2

This is a continuation of Case 00-18394, having made a full circle from Broward (trial court, 11/2000) to the Appellate Court (2001), then Administrative Court (2001) then back to Broward (2002) after the legislature embedded the 1900-ft policy into Florida's laws in 2002 and created a new type of warrant called "agriculture warrants." The Department also modified their rules to be consistent with the new statutes.

It is ironic that the Department argued in the Appellate Court that the case had to be heard first in Administrative Court. Then, as the trial began in Administrative Court with depositions, the Department was able to convince the legislature to change the law, and embed the 1900-ft policy in the Florida statutes. This resulted in the case returning again to the Broward District Court. The Department seemed to be playing a "cat and mouse" game, to avoid a trial on the merits of the 1900-ft policy.

Issues Involved: Fair compensation for cutting healthy trees, entry into properties using blanket warrants as provided for under the "new canker laws." It was also contended that the 1900-ft policy was unsupported by reliable science.

Final Outcome: In May 24, 2002, Broward Court Judge ruled in favor of Plaintiffs, and for a second time, and healthy tree cutting was stopped. It was a temporary injunction.

On January 15, 2003 the Appellate Court reversed Judge Fleet's ruling on the issue of science and requirement of fair compensation, however let stand the requirement for search warrants. They permitted multiple addresses on the same warrant. The issue of compensation and level of scrutiny allowed in examining the 1900-ft rule went on to the Florida Supreme Court.

The major delay in healthy tree cutting came as the Department chose to appeal the Broward Court decision, rather than provide additional support for their 1900-ft rule and offer fair compensation to residents. The Department can always have an injunction lifted by complying with its objections.

Details: Plaintiffs in this case were Broward and Miami-Dade Counties with other municipalities joining the lawsuit. Trial lasted approximately three weeks. Dr. Gottwald testified for several days on the scientific merits of the program. The municipalities provided expert witnesses in the general area of epidemiology and statistical analysis, but no expert in both citrus canker and plant epidemiology.

The opinion handed down by Judge Fleet on May 24, 2002 states the following:

- The 2002 amendments to FS 581.184 and FS 933.02 are in direct contravention of Article I Section 12 of Florida's Constitution and the Fourth Amendment of the US Constitution and are not enforceable.

- The Science upon which such amendments are predicated is significantly unsound and therefore is not constitutionally acceptable as a basis for legislative abrogation of a property owner's right to the full panoply of protections afforded by our State and Federal constitutions.

- All citrus trees not patently demonstrating the existence of citrus canker pathogens have a determinable value and cannot be destroyed by the state in the absence of full and fair compensation determined by appropriate condemnation proceedings. This applies to commercial groves, citrus trees owned by municipal governments and citrus trees owned by private parties.

His order went on to require search warrants for entry onto property. The search warrant must specify the address. Blanket agricultural warrants were unacceptable. Since Judge Fleet ruled on motion for temporary injunctive relief, which is usually a request to order a halt to an activity pending a trial for permanent relief, the county attorneys wanted to go back to Broward Court to challenge the epidemiology study at a deeper level (full disclosure of all collected data would be required). The issue of "level of scrutiny" and compensation would be determined by the Florida Supreme Court.

The Department wanted the case to be immediately transferred to the Florida Supreme Court for a decision. The Fourth DCA agreed on July 9, 2002 for immediate transfer (Case 4D02-2584) but the Supreme Court did not agree to the immediate transfer.

On January 15, 2003, the Fourth District Court of Appeals reversed a number of findings in Judge Fleet's decision. The Court reversed Judge Fleet in his determination that the "citrus canker law" as passed in year 2002, FS 581.184, violated substantive and procedural due process. However, the court upheld Fleet's finding that "blanket agricultural search warrants" per FS 933.07(2) were in violation of Fourth Amendment rights of unreasonable search. However, the Department was permitted to place multiple addresses on a single warrant, and use electronic signatures.

The Corneal decision, was considered key to the Fourth DCA opinion as follows,

> In *Corneal*, the court considered whether these regulations requiring destruction of healthy trees without compensation was a constitutional use of the state's police power. Acknowledging that the state's police power to protect the public welfare is very broad, it is concluded that :the absolute destruction of property is an extreme exercise of the police power and is justified only within the narrowest limits of actual necessity, *unless the state choses to pay compensation.*

There was at this point, no question if compensation was being paid to residents, at this was passed into law as FS 581.1845 in year 2002. Now, by understanding the Corneal decision, it was clear why FDACS had such a quick change of heart on paying resident's compensation after years of declaring the exposed trees were worthless.

Broward County Case 2 decided by the Florida Supreme Court, February 14, 2004

Issues: While a number of issues were before the court, the court focused on the level of scrutiny in examining the 1900-ft policy given there was compensation for healthy trees as established by the law passed in 2002. It was argued that the lawsuits for additional compensation would be difficult because it would require an act of the legislature to appropriate the funds. The Department argued the trees within the eradication circle were worthless and constituted a public nuisance. The plaintiffs wanted to return to Broward Circuit Court for a permanent injunction. This would likely have entailed a more detailed examination of the field study to determine if the 1900-ft policy was within the narrow limits of necessity to eradicate citrus canker.

Final Outcome: The Supreme Court ruled a lower standard of court scrutiny of eradication policy is appropriate (reasonable relationship) when compensation is offered. Under this standard, all that needs to be shown was that a relation existed between their actions and what they intended to accomplish. They cite the case *Lane v. Chiles*, "A statute is valid if it bears a rational relation to a legitimate legislative purpose of safeguarding the public health, safety or general welfare and is not discriminatory, arbitrary or oppressive."

The Florida Supreme Court states (page 13 of the opinion), "Because the Fourth District concluded that the Citrus Canker Law is a valid exercise of the State's police power and provides for compensation for the destruction of trees having value, the court determined the reasonable relationship test applied. "

On page 19, the court states:

> In this case, we conclude that under the statutory scheme, the State is obligated to provide more than token compensation if the State has destroyed a healthy, albeit exposed tree.

Details: This ruling ended the further legal challenges to the science issue. The Court's ruling stated exposed trees have value, but the lingering question is should they be valued as much as healthy trees outside of the eradication circle.

The Court indicated that "token compensation" is not acceptable, but how does one differentiate between "token compensation" to "full compensation" if the trees within the circles are more likely to become diseased as per the Department? This argument would persist for years through the class action lawsuits.

The attorney for Broward County, Mr. Andrew Meyers stated during oral arguments that residents could not go to court if they felt the compensation was inadequate, because (a) the Department would fight "tooth and nail" against paying one cent more than what it was legally required and (b) even if residents prevailed in obtaining full compensation, it would require legislative action to be able to collect. These two events have occurred just as Attorney Meyers had predicted.

The ruling by the Supreme Court seems to lack practicality. Had the Department been required to set aside funds to cover future compensation claims, and a neutral third party to assess compensation claims, it is certain that many of the legal problems could have been avoided.

7. Broward County Case 3 (Predates Florida Supreme Court Decision)

Issues Involved: Broward, Miami-Dade and other municipalities charged the Department had inaccurate testing for canker. Errors in testing would result in unnecessary cutting of healthy trees in violation of State laws. Further it was charged that the FDACS procedure in delineating the 1900-ft zones was not in accordance to the law. Finally, the Department had not made adequate provisions for compensating residents for loss of their healthy trees.

Final Outcome: Judge Fleet ruled on July 18, 2003 that testing and measurement procedures were inadequate. He ruled that compensation was required. However, his rulings were reversed on appeals as the Circuit Court ruled the Broward County court was the inappropriate venue to hear these administrative policy complaints.

Details: The Florida Law 581.184 required that all citrus trees within 1900-ft of an infected tree be removed. Measurement were made from the centroid point of the lot with the infected tree to the centroid of the lot with healthy trees. This was done for convenience, as the location coordinates defining the property boundaries were stored in their computer. The error can be large if the lots are large. The Department reduced the radius to 1851-ft, to give the benefit of error to residents. However, the Department did not have the authority to make allowances, as the 1900-ft policy was part of the law. Further, the Department had taken the position in its successful appeal to the Fourth DCA that any radius less than 1900-ft would not achieve their goal of canker eradication. So, ironically, the Department made the 1900-ft policy a rigid, absolute policy to win the appeal of Broward Case 2 and lost this case as a result of it.

The court concluded that the Department had used unreliable tests to determine if a tree was infected with Asian strain citrus canker. It is noted in its opinion, the Wellington strain is only considered a less virulent strain of citrus canker. The ruling did not go far enough into the problem. As discussed in Chapter 3, this strain is known to infect only lime and alemow trees, and the latter cultivar is rarely found in Florida. The ruling by Judge Fleet, makes it clear that the testing which is reliable in the "vast majority of cases" is not sufficient when a single infected tree can result in destruction of 260 acres.

8. Broward Cases Results Summary

Broward Case 1 resulted in an injunction against the cutting of healthy trees only within residential properties of Broward County. Approximately 134,000 healthy trees were destroyed in Broward County as a result of the 1900-ft rule as compared to 16 million trees for the program. (Source: www.citruscankerlitigation.com)

The passage of the new canker laws in year 2002, did not result in the intended consequences of relieving the Department of lawsuits by the municipalities and residents of South Florida. In fact, the new laws established the Broward District Court as the proper venue to challenge the legality of the 1900-ft rule, as opposed to administrative court. It was only a stop gap measure to stop the ongoing case in Administrative Court.

Since Canker War II, the mantra of the Department was exposed trees are worthless trees. Then in year 2002, the Department seemed suddenly to acknowledge exposed trees had value with passage of the new canker law. But, this was just a legal maneuver, as this would limit scrutiny of the 1900-ft policy of the Broward Court. And it worked, even though the new canker law never recognized compensation was for replacement of a healthy tree nor based on the value of the healthy tree. The ruling by the Florida Supreme Court established that since compensation in this limited way had been made, court scrutiny of the program was limited to a rational basis, and not within the narrowest limits of necessity. With this ruling, there were no further legal challenges to the 1900-ft policy.

Probably the most enduring contribution of these lawsuits for residents was the Department must use warrants when entering residential properties without homeowner's consent. The Department lost the argument that they were in essence in hot pursuit, when chasing citrus canker. The courts also established that cutting down healthy citrus trees was a "taking" and required compensation.

The Department argued that the destruction of private property is necessary when it is determined to be a public nuisance by being exposed to a disease, and they alone were the arbiters of the extent of the exposure zone. Although the court ultimately rejected this argument, it is a lesson learned that any legal challenge can be costly and extremely time consuming to contest the self proclaimed powers of the Department, but it can be done.

A final lesson is how incredibly difficult it is to file any lawsuit against a government entity. There is a process of legal condemnation of property for public use and just compensation for homeowners. It is used all the time, in public construction projects. But, the Department felt it had sufficient rights to declare all healthy trees within a 1900-ft radius circle to be worthless. The Supreme Court ruled that these trees have value. The compensation cases, begun in year 2000 have not been resolved. The Appellate court seemed to support an indefinitely delay in payments to class action members, awaiting the legislature appropriates the funds. The right to just compensation seems only a distant promise to residents.

9. FDACS Documents Released as a Result of Court Cases

1. Dr. Gottwald made a presentation in the Broward Court on November 8, 2000. Copies of many of these viewgraphs are provided in the online support website. They appear to be the same viewgraphs as presented at the June 2000 International Citrus Canker Workshop as the conference name is in the footnote of every slide. The information given on the viewgraphs seems to coincide in many cases with the Workshop transcripts. The same viewgraphs may have been used in both Broward and Workshop presentations, but FDACS has not confirmed this. The USDA ignored my request for workshop presentation slides.

2: Key correspondence between Dr. Gottwald and FDACS prior to the enactment of the 1900 ft rule. Attorneys refer to one document as the "Interim Report."

3. A report from the May 11, 1999 risk assessment group meeting.

4. Two additional articles on citrus canker as published in 1997, one in the trade journal, Fruits, and the second in Citrus Industry journal.

5. A manuscript for a Letter to the Editor article to appear in Phytopathology with Dr. Gottwald, as lead author was submitted to the Broward court along with a letter from the Editor in Chief of Phytopathology, indicating the letter had been approved for publication.

From the Miami-Dade case, a list of 2700 infected trees with addresses was produced. This would prove to be an important piece of evidence in how the eradication program was proceeding. This information has been posted to the online supporting documents website.

10. Other Cases Seeking Injunctive Relief

Miami-Dade Case

Issues involved: Miami-Dade County and the City of North Miami filed lawsuit against the program for a preliminary injunction on the basis that search warrants were required for inspection and tree destruction.

Final Outcome: Judge Friedman in May 18, 2001 ruled that search warrants were required, but stated that mass-produced "rubber stamped' warrants would be acceptable so the program would not be impeded. His ruling was overturned by the Third District Court of Appeals based on the fact there was no immediate threat of injury to the parties as they did not own citrus trees.

Details: In reviewing the exceptions to search warrants as cited by the Department, Judge Friedman opined, "There are no cases on 'all fours' [four exceptions to the requirement for warrants cited by defendants]. At the time of the decision (18-Apr-2001), healthy tree cutting had been stopped by the Department, awaiting a decision by the 4th DCA on the Broward Case 1.

Judge Freidman stated:

> While this Court finds *Camara* case to be legally compelling (requiring warrants), the practical side of this Court says that warrant requirements are much ado about nothing. The necessity of obtaining warrants under the facts of the CCEP is actually a waste of judicial effort, State time and money. ... Therefore, issuance of warrants will be nothing more than a rubber stamp.

Friedman's ruling seems contradictory with a finding that the search without warrants would be an unreasonable intrusion but mass produced warrants would not provide any more rights to residents, just a way around Constitutional guarantees.

The ruling was very important, because for the first time, a judge was stating that entry onto properties was a violation of Florida and United States constitutional rights which no rule or law could remedy.

In the Broward case, the attorneys for the plaintiffs would not make the same mistake and include residents with citrus trees that had been issued IFO's but not yet cut down. However, this challenge would not come until May 2002, after the Florida legislature had passed new laws in support of the CCEP.

Dr. Fred Westphal and C.A.D.E.T Case (January 2001)

Issues involved: Dr. Fred Westphal, a retired professor of philosophy at the University of Miami, filed a lawsuit in October 2000, on behalf of himself and an organization he founded, called Citizens Against the Destruction and Eradication of Trees (C.A.D.E.T.) asserting that the CCEP had violated the constitutional rights of individuals, including the protection against unreasonable searches. A "Science Review" meeting was held in December 2000 with the staff scientists with FDACS/DPI.

Final Outcome: Judge Freidman dismissed the case in January 2001, citing that there had to be a pending action of eradication against the Plaintiffs. In this case, their trees had already been cut down.

City of Delray Beach Case, 01-8273-CIV-FERGUSON, US District Court, Southern District of Florida (July 30, 2001)

Issues involved: Arguments presented against cutting of healthy trees were similar to Broward Case 2, a) Violation of constitutional rights of unreasonable searches, b) No compensation and c) Due process and equal protection violations by the Department. An evidentiary trial was held in which Drs. Gabriel (Head, Plant Pathology Dept, UF) and Graham (Professor of plant pathology, UF/IFAS) testified that a new chemical Oxidate is not an effective treatment citrus canker.

Final Outcome: On July 30, 2001, Judge Ferguson dismissed the motion for injunctive relief stating the case had to be heard in State courts rather than Federal court. City of Delray Beach attorney Jay Jambeck and private attorney Barry Silver represented the city and homeowners.

11. Compensation Cases (Inverse Condemnation Cases)

The first class-action lawsuit to be certified occurred in 1997 (Varela Case) when cutting was limited to a 125-ft radius zone. It was the dismissed in 1999 by the 3rd District Court of Appeals based on the Polk case.

The class-action lawsuits filed after enactment of the 1900-ft rule have been more successful, as cases have gone to jury trials in several counties. These case seek additional compensation for residents who lost their healthy citrus trees. Municipal and County legal offices were not involved in any of these cases. Lead attorney for these cases is Robert Gilbert, a private attorney. Filings of lawsuits were delayed pending the result of *Patchen v. FDACS*, which posed a question that the 3rd District Court deemed of great public importance to be answered by the Florida Supreme Court (to paraphrase a bit) as: "Did a prior Florida Supreme Court ruling that the value of commercial trees located within 125 ft zone were worthless (no compensation) be extended to backyard trees in a 1900 ft zone?" The prior Supreme Court decision is *Department of Agriculture & Consumer Services v. Polk*, 568 So. 2d 25 (Fla 1990).

The Florida Supreme Court opined that backyard citrus within the 1900-ft zone had value. But establishing a value for the healthy trees (the Department favored calling these trees "exposed trees") would be a contentious issue in the lawsuit, as the Department tried to introduce into evidence that these trees had little or no value, because they were destined to be diseased.

Under the Shade Florida Program, the residents that lost citrus trees are given a debit card they can use at Walmart which provides $100 credit for the first citrus tree removed, and $55 for the each citrus tree after the first one.

It is noted that class action lawsuits for damages are very common, particularly with publically traded companies which have failed to make full disclosures or mislead investors in their annual financial report or other communication. In these lawsuits, sometimes as many as 20 law firms compete for the opportunity to represent the class, by searching for individuals who can demonstrate the greatest damages, typically those with large equity holdings in the company being sued.

In stark contrast, there was no obvious competition for class representation in the case against the Department. The compensation case appears to be one that many attorneys shun due to the many impediments of being able to present a case in front of a jury and then collecting the additional compensation for the class.

Note, that there is nothing in the US Constitution giving immunity to the governmental agencies from the Fifth Amendment (right against taking of private property without compensation) and to Article 1, Section 12 of the Florida Constitution. Further, no laws or rules can be made by States that limit rights within the US Constitution.

An estimate of the value of a citrus tree had been prepared in 1999, by Connie Riherd working with FDACS/DPI in a memo to Richard Gaskalla identifying the value of healthy citrus trees to be $468. This memo, the 18-page Florida Supreme Court opinion and other legal documents are provided in the online support documents website.

Additional Compensation Awarded

In the Florida Supreme Court Case challenging the legality of the CCEP (decision on Feb 12, 2004), the Department argued vigorously that any resident who was dissatisfied with the compensation could seek remedy through the judicial system. However, Andrew Meyers, representing Broward County, told the justices of the Florida Supreme Court during oral arguments that the Department would "fight tooth and nail" before paying one cent more than they had to by law.

The legal process has proceeded along these 4 steps: (1) Hearing to determine if class can be certified, (2) Class goes before Judge to show Department is, in fact, liable for damages, (3) Jury trial for determination of compensation amount followed by entry of final judgment and interests and (4) Trial court also awards attorney's fees and costs.

At each step, the decisions of the court can be appealed to the District Court of Appeals. The Department appealed the Broward case, which the Court found baseless, "We find the Department's arguments to be frivolous. No matter how one looks at the facts, the owners prevailed on the significant issues." Fourth District Court of Appeals, No 4D09-979, January 26, 2011.

Despite these impediments, the trial court granted additional compensation as follows:

Palm Beach County:	19.0 million dollars,	based on trial in Mar 2015
Broward County:	9.0 million dollars	
Orange County :	20.6 million dollars	
Lee County:	9.6 million dollars	

The Broward court claim is approximately 15.0 million dollars, today as interest and legal fees are added to the total. Links are provided on the online website, to these cases and will be periodically updated under the Legal- Compensation tab.

12. Florida Refuses to Pay Tree Owners

The following news story was posted on the Internet, on the CBS News Miami website on September 25, 2011. A link is maintained to the full story on the online supporting documents website:

After waiting for more than a decade for the state to pay up, homeowners in two South Florida counties whose citrus trees were chopped down by the state to stop the spread of citrus canker are still for compensation.

"I just think it's a new low," said Tim Farley whose orange, grapefruit and lemon trees were cut down. "They did an injustice to the people and they have to pay us. They are spending millions of dollars of taxpayer money to fight us. They are slapping the people like they always have."

Like thousands of others, Farley told the Associated Press he was dismayed when the state chopped down his backyard trees, but at least he figured he'd get paid after judges ordered Florida years ago to compensate him and others.

Yet despite rulings in favor of tree owners in two counties, the state Department of Agriculture and Consumer Services has so far refused to pay Farley and many others so much as a nickel.

The state's position is that the Legislature would have to approve a special bill — known as a claims bill — before any money can be paid. So far, the decisions in Broward and Palm Beach counties have the state on the hook for about $30 million in combined compensation. There are still cases pending in Miami-Dade, Orange and Lee counties that could account for millions more.

"All the money comes from taxpayers," said Wes Parsons, an attorney representing the Agriculture Department. "If they are paid, the Legislature must appropriate the money to do so. There has not been a claims bill filed."

Attorney Robert Gilbert, who represents the former tree owners, said it shouldn't come to that. Gilbert said judges have ruled that the destruction of the trees amounted to a government "taking" that requires more immediate fair compensation — similar to when the government condemns private property to put a road through.

The 4th District Court of Appeal, in upholding the Broward County decision in favor of tree owners, rejected the state's argument that the trees were a "public nuisance" that had no value because of the potential they could spread the disease. Instead, the judges said, the trees had value and were chopped down mainly to benefit Florida's citrus industry.

"If trees are destroyed not to prevent harm but instead to benefit an industry, it is difficult to understand how (the Agriculture Department) can argue on appeal that the trees legally constituted a nuisance without any value," the judges ruled in May 2010. "If government cuts down and burns private property having value, then government has taken it. And if government has taken it, government must pay for it."

The state did give homeowners a $100 debit card good at Wal-Mart garden centers for the first tree that was removed and $55 cash for tree after that. The tree replacement programs, which cost the state $44 million, ended in 2009. But Farley and thousands of other tree owners considered that too little for healthy trees that in some cases were 30 feet tall, so they filed lawsuits representing classes of all affected tree owners.

Legal battles over control of agricultural diseases and pests go back many decades nationally. The landmark U.S. Supreme Court ruling on the subject came in 1928, when the justices upheld Virginia's right to cut down a grove of red cedar trees to protect nearby apple orchards from rust disease.

The justices found that the Constitution allows states to carry out "the destruction of one class of property in order to save another which, in the judgment of the legislature, is of greater value to the public." But that ruling did not address the subject of fair compensation.

In Florida, the healthy backyard trees were cut down and burned during the state's decade-long, $623 million attempt to eradicate citrus canker. The fight was ultimately abandoned in 2006 after a series of hurricanes, particularly Hurricane Wilma in 2005, spread the wind-borne disease widely across the state.

Canker causes blemishes and can make fruit drop prematurely, but the disease does not kill infected trees and ripe fruit can still be eaten or made into juice. Still, the disease is considered a major threat to Florida's $9 billion citrus industry, the largest in the U.S. The canker bacteria doesn't affect humans.

It was first identified in Florida in 1910, declared eradicated in 1933, but then appeared again in 1986. The latest war on canker started in 1994. Beginning in 2000, the state destroyed any residential tree within a 1,900-foot radius of an infected tree — even if there was no evidence of infection on the trees earmarked for destruction.

All told, more than 16.5 million trees were destroyed, including more than 865,000 residential trees. The rest were at commercial groves and nurseries.

In Broward County, a jury authorized an additional $34 per tree and the Palm Beach case would pay $210 more for each tree. The state continues to appeal both rulings, and ultimately the overall issue of additional compensation is likely to land at the state Supreme Court.

Gilbert contends the time has long passed for payments to be made and said he will return to court in an effort to force the state to compensate those who lost trees, rather than relying on the whims of the Legislature.

In July 2012, the Fourth DCA, rejected a request from the Broward and Palm Beach Class to collect compensation from the Department as determined in the trial. The basis for this rejection was the request was pre-mature and the class representing home owners must work through the appropriations process.

The latest appeal by the Fourth District Court of Appeals opinion on May 4, 2016 appears to further delay payment from the FDACS. The court decided the class must first attempt to obtain a "writ of mandamus" to secure payment. However, without the legislature cooperation in approving an appropriation amendment, this attempt seems doomed. The court indicated that it would likely take up the question of the constitutionality of laws designed to shield government agencies from class action lawsuits if the writ was denied.

Thus, suing any governmental agency for damages is either extremely protracted time consuming process or just plain impossible unless you have the judiciary, legislature and executive branches are all on your side. It would appear that the constitutional question needs to be resolved at the Florida Supreme Court level so it would be binding on all courts in the state.

The outcome of the class action lawsuits will be posted on the website, www.citruscankerdocs.com.

13. Concluding Remarks

The Broward cases (*Haire v. Florida Department of Agriculture)* extend far beyond the cutting of citrus trees. It redeemed basic rights of residents against unreasonable searches. It further put the Department on notice that just compensation was required for destroyed healthy trees These rights will be applicable to any future actions of government agencies.

The Department defended its action as necessary to safeguard Florida against a menacing disease that was threatening the public welfare of Florida. Just as the federal government has taken extraordinary steps in time of war, so the Department of Agriculture felt its actions were defendable given the threat citrus canker to its citrus industry.

Since funds had been appropriated by the legislature and signed by the governor, the only door open to challenge the program was the courts, Only with the collaborative efforts of attorneys from Broward and Miami-Dade counties and 8 municipalities could residents bring these issues into a legal forum. Otherwise, residents were basically powerless to navigate through the legal system, and bring suit against the Department.

To residents, it should be the burden of the Department to demonstrate the necessity of the 1900-ft rule. Eradication programs had to respect the constitutional guarantees against unreasonable searches, and seizure of property without compensation. From their perspective, the Department can not be above the law. The Department can not implement a program without legally required rule making procedures which includes public hearings,

The Department stated in various public forums, that the legal actions of a few residents only made the epidemic worse, causing many more trees to be destroyed. Even worse, is the idea that a few residents impeded the program to such an extent, as to cause it to fail.

The Department has never conceded there was any mistakes made, although the courts ruled against them three times in Broward court and once in Administrative court. These cases involved only the destruction of healthy residential trees.

The passage of the new canker laws in year 2002, did not result in the intended consequences of relieving the Department of lawsuits by the municipalities and residents of South Florida. In fact, the new laws established the Broward District Court as the proper venue to challenge the legality of the 1900-ft rule, as opposed to administrative court. It was only a stop gap measure to stop the ongoing case in Administrative Court.

Since Canker War II, the mantra of the Department was exposed trees are worthless trees. Then in year 2002, the Department seemed suddenly to acknowledge exposed trees had value with passage of the new canker law. But, this was just a legal maneuver, as this would limit scrutiny of the 1900-ft policy of the Broward Court. And it worked, even though the new canker law never recognized compensation was for replacement of a healthy tree nor based on the value of the healthy tree. The ruling by the Florida Supreme Court established that since compensation was provided, albeit limited, court scrutiny of the program was limited to a rational basis, and not on a strict scrutiny basis, which is within the narrowest limits of necessity. With this ruling, there were no viable legal challenges to the basis of the 1900-ft policy.

Probably the most enduring contributions of these lawsuits for residents is the Department must use warrants when entering residential properties without homeowner's consent. The Department lost the argument that they were in essence in hot pursuit, when chasing citrus canker. The courts also established that cutting down healthy citrus trees was a "taking" and required compensation.

The Department argued that the destruction of private property is necessary when it is determined to be a public nuisance by being exposed to a disease, and they alone were the arbiters of the extent of the exposure zone. Although the court ultimately rejected this argument, it is a lesson learned that any legal challenge can be costly and extremely time consuming to contest the self proclaimed powers of the Department, but that it can be done.

There is a process of legal condemnation of property for public use and just compensation for homeowners. It is used all the time in public construction projects. However, the Department felt it had sufficient rights to declare all healthy trees within a 1900-ft radius circle to be worthless. The Supreme Court ruled that these trees have value. The compensation cases, begun in year 2000 have not been resolved. The Appellate court seemed to support an indefinitely delay in payments to class action members, awaiting the legislature appropriates the funds. The right to just compensation seems only a distant promise to residents.

Short Notes on the Website

SN 1.6 What is an exposed tree?
SN 3.1 The Case of the Dirty Scalpel
SN 5.1 Peer Review of the Field Study
SN 5.2 Broward Attorney Andrew Meyers' Comments
SN 5.3 The Unfinished Trial of the Citrus Canker Eradication Program

Florida Statutes:

1984 FS 581.184 – Authorizing eradication of citrus plants
2000 FS 581.184 – Revised for Canker Wars III, giving FDACS additional authority to destroy citrus which harbored canker but whose symptoms had not yet appeared.
2002 FS 581.184 – Revised to avoid legal challenges, now the law simply gave FDACS authority to destroy citrus which was located within 1900 ft of an infected tree.
2002 FS 582.1845- Amendment providing cash compensation to residents for destroyed citrus trees.
2002 FS 933 – Amended to include special agricultural warrants that would be county wide searches of backyards on the basis that canker had been discovered on one property in the county. In Miami-Dade county, this amendment would give inspectors and cutting crews authority to enter all residents in a 1000 square mile area. Ruled unconstitutional in the 4th District Court of Appeals as a violation of the Florida and US Constitution giving protection against unreasonable searches.

Attorneys:

Attorney Andrew Meyers representing Broward County was the lead attorney on all Broward cases and appeals. Attorney Robert Duvall representing Miami-Dade County was the lead attorney on the Miami-Dade Case and in Broward Case 2, where the unauthorized entry onto properties was an issue. Other attorneys include:

R. Ginsberg , Miami-Dade County, Earl Gallop, representing North Miami in Miami-Dade Case E. Dion and T. Scrudders attorneys for Broward County, C. Kalil and M. Misuraca, private attorneys, Private attorney Barry Silver also participated in Broward Case 2 at no charge

In arguing many of the cases in the trial court for FDACS, Mr. Wesley Parsons from Adorno and Zeder, PA represented the Department. Many attorneys from Greenberg Traurig assisted the Department including Attorney Arthur England who argued their case in the Florida Supreme Court, Elliot Scherker, Elliot Kula, Jerold Budney, and David Ashburn. Broward Case 1: (Nov 2001- Mar, 2002)

Broward Case 1:

Nov 17, 2001: Broward Circuit Court, Granting temporary injunction, November 17, 2001, FDACS vs City of Pompano. Case 00-18394 (08) CACE. Opinion by Judge Fleet.
July 11, 2001: Fourth District Court of Appeals reverses trial court decision, Case 4D00-4116. Electronic copy, pdf.
July 31, 2001: Administrative Hearing rules against FDACS, re-affirms requirement of rule making for 1900 ft rule, July 31, 2001, Judge Laningham. Case 00-4520RX. DOAH. Electronic copy, pdf.
Aug 17, 2001: 1st District Court of Appeals, Department appeal to 1st DCA is rejected, rule making must proceed.
Nov 21, 2001: Briefs filed challenging new rule in Administrative court. Pre-filing electronic copy available of administrative complaint, pdf.

Mar 18, 2002: Citrus canker law, FS 581.184 and agriculture blanket warrants, FS 933.02 signed into law by Governor Bush. Administrative court legal challenge is rendered moot.

Broward Case 2: (May 2002 to Feb 12, 2004)

May 24, 2002: Broward Circuit Court. Temporary Injunctive Relief Granted, Haire et al vs FDACS, Judge Fleet. Case 00-18394. Rules that both FS 581.184 and FS 933.02 are unconstitutional.
Jul 9, 2002: Fourth District Court of Appeals approves immediate transfer to Florida Supreme Court, however Supreme Court returns case to 4th DCA.
Jan 15, 2003: Fourth District Court of Appeals partially reverses decision. Court rules FS 581.184 is constitutional. Affirms blanket warrants are unconstitutional, but permits electronic signature and multiple names/addresses on single warrant. Opinion written by Judge Warner. Electronic copy available
February 12, 2004: Florida Supreme Court decision. Rational relationship standard holds since compensation was provided. Considers compensation sets a "floor" value. Warrants upheld.

Broward Case 3, Granting Temporary Injunction

July 18, 2003, *FDACS vs. Haire*, Case 00-18394(08), Judge Fleet. Grants injunction based on improper laboratory determination of canker and measurement issues. Reversed on appeal, January 21, 2004 by the Fourth District Court of Appeals stating these issues need to be first decided in Administrative Court.

Miami-Dade Case:: Granting Temporary Injunction, April 18, 2001, Miami-Dade County vs. FDACS, 11th Judicial Circuit. 00-30193 CA 02. Judge Friedman. Reversed on Appeal.

Dr. Westphal case: Complaint for Injunctive Relief, filed on October 30, 2000, *Fred Westphal v. Jeb Bush and Bob Crawford*, Case 00-28425 CA 02, 11th Judicial Court. Judge Freidman. Case not appealed.

Online legal documents are provided on the supporting documents website:

www.citruscankerdocs.com

6. SELECTED TOPICS IN EPIDEMIOLOGY

> We live in a society exquisitely dependent on science and technology, in which hardly anyone knows anything about science and technology.

Carl Sagan

> Lest men suspect your tale untrue, Keep probability in view.

John Gay

---

1. Introduction

This chapter provides basic terms and concepts of epidemiology in relation to the citrus canker disease. Every effort has been made to keep the focus on concepts rather than a technical or mathematical discussion. Recent experimental studies providing transport distance of citrus canker bacteria under simulated windblown conditions are summarized at the end of this chapter.

While the topics presented are useful in understanding the Florida field study, no review of the field study is provided in this chapter. The review of the field study is presented in detail in the appendices. These findings are then summarized in Chapter 7 in this book. Thus, it is hoped that the intrepid reader will finish this chapter, then continue with the online appendices before rejoining the discussion in the book in Chapter 7. Admittedly, this combination of both print and online information is unusual, but it was the best means of providing all essential information to the reader.

Commissioner Crawford on February 26, 1998 requested a 12 month field study of citrus canker in 3 selected residential areas within Miami-Dade County. Within these sites, there would be repeated inspections using a unique form and the location of all citrus trees were identified.[13,14] It is also assumed that the same inspectors would fill out the normal inspection forms used by the CCEP and this information would be entered into the CCEP database.

The field study was unusual, because it was conducted in residential areas instead of commercial settings or experimental plots. The unique challenges of an observational study in residential areas are discussed in this chapter.

It is noted that the field study is referred to by FDACS as the "Epidemiology Study." However, within this book, it is always referred to as the "field study" to avoid any ambiguity between the study based on the site information and other related research on citrus canker.

2. Key References

Two excellent textbooks on plant disease epidemiology were invaluable to the preparation of this chapter as authored by Campbell and Madden published in 1990 [4] and Madden, Hughes, and van den Bosch published in 2007.[19] Topics in plant disease epidemiology from Campbell and Madden's book include:

- Monitoring epidemics: host, environment, pathogen
- Modeling and data analysis
- Temporal analysis of epidemics: Description and comparison of disease progress curves, advanced topics
- Spatial aspects of plant disease: Dispersal gradients, long-range transport, analysis of spatial patterns
- Simulation models of plant disease
- Designing experiments and sampling

As this chapter is narrowly focused on citrus canker epidemiology, the key references would be those which the Department has claimed were critical sources of information on setting eradication policy.

The Department has claimed the 125-ft policy is supported by the observational study conducted by Dr. Stall in Argentina from 1978 to 1979. The results of the study were published as a research note in 1980 providing few details.[25] A complete report on the study was never published, thus the 125-ft policy does not appear to be well supported.

The Department has also considered the two articles on the Florida field study to be foundational documents to their 1900-ft policy. The field study is documented in an article published in April 2002 in Phytopathology journal. This article is available free of charge from the American Phytopathological Society (APS) website at www.apsnet.org and on the online supporting documents website. Other supporting documents on the field study are listed in the appendices.

3. Plant Disease Epidemiology

A definition of plant disease epidemiology as given in Campbell[4] is, "The study of the temporal and spatial changes that occur during epidemics of plant diseases that are by populations of pathogens in population of plants."

The plants can be part of any agricultural, horticultural, forest or natural ecosystem. The discipline of plant disease epidemiology is not specifically tasked with eradication of an epidemic. Most plant diseases are very difficult to eradicate, as stated in Madden[14] , page 1:

> Unlike the situation with several highly contagious animal diseases, very few plant diseases have been eradicated, hence some inoculum of the pathogen can usually be found in a given area for the important diseases of a given crop.

The authors then state that a typical disease management goal is to maintain a low intensity of the disease, rather than total eradication. Additional discussion of control verses management options are provided in the online supporting documents website.

A valid epidemiology study may in some way, simply advance the understanding of a disease, rather than recommend a means to control or eliminate it. This understanding may contribute to the selecting the best strategy, along with many other considerations, such as legal requirements, finance and logistics.

Plant disease epidemiology is a scientific discipline which has evolved greatly in the last several decades. Mathematical modeling and statistical analyses are an integral part of plant disease epidemiology. The discipline has made great advances by borrowing and adapting techniques from a multitude of other disciplines, including:

- Descriptive and qualitative statistical analyses
- Ecological and environmental modeling
- System analyses, modeling and optimization in the discipline of operations research

Evaluation of epidemics is done according to the level of knowledge of the pathosystem (all elements contributing to the development of the disease) often using tools adapted from other disciplines, including botany, ecology, genetics, physics and statistics.[14]

Assessments can be done through experimental and observational studies. If experimental and observation studies show different results, then perhaps a re-examination of the underlying description and assumptions is in order. This is true of most research studies where practical.

Most epidemiology studies involving plant disease in a commercial setting, i.e. nursery, farms or experimental plots and not residential settings. Even with a fairly uniform host environment, quantifying epidemic behaviors is difficult. An early pioneer in epidemiology, Ernst Gaumann wrote in 1946 (Campbell[4], page 16):

> Every epidemic develops by its own rules, changes its character, expands and becomes malignant, decreases and becomes milder: it has an appearance of its own, its own morphology, its own *genius epidemicus.*

The authors further explain, "The *genius epidemicus* identifies the typical characteristics that distinguish one epidemic caused by a specific pathogen on a host from other epidemics caused by a specific pathogen on the same host. "

The collection of more data does not necessarily mean better or more definitive results are obtained if the mechanisms of dissemination are changing during data collection. As discussed in the appendices, unique conditions existing within residential areas, add considerable complexity which may compromise observational studies or render certain statistical analyses meaningless. It would be no fault of researchers to conclude the data are too compromised to apply certain methods or make valid conclusions.

Statistics is an integral part of epidemiology, which includes procedures for proper data collection and sufficient controls in place to avoid errors and outside influences. Studies should be well planned in well in advance and any change during the course of the study should be well explained. The validity of results is dependent on appropriateness of a statistical procedure. The exclusion of certain data may bias or invalidate certain results.

4. Terms and Concepts Related to Field Surveys

The selected terms and concepts in this section are directly related to epidemiology studies of citrus canker in Florida. A survey involves collection of data, which can be categorized as experimental or observational in nature as follows:[27]

> **Experimental study:** Some treatment is applied to the subject or object under study, then we observe the effect on this subject.
>
> **Observational study:** Specific characteristics are observed and measured. There is no attempt to modify the subjects being studied.

In terms of experimental study to study dispersion or dissemination of a pathogen, the typical "treatment" consists of introducing one or more diseased plants (source or foci of disease) into a group of health plants. As discussed further in the section of experiments conducted with citrus canker, additional "treatments" may consist of measures to limit canker dissemination, such as applying bactericides or employing wind breaks. In these cases, it is necessary to have a comparative case, or control experiment, without these treatments.

Experiments in epidemiology tend to show variation because of the variability of biological processes, no matter how well controlled the external factors. Studies may require a specialist in statistical analysis in analyzed the variation of experimental results.

Experimental studies have an advantage over observational studies as high level of control and isolation can be imposed. Both observational and experimental studies may be useful the understanding of processes. If conclusions based on observational studies are confirmed with experimental studies, this furthers the credibility of the conclusions. Also, if the outcomes of the experiment can be explained by the current knowledge of underlying mechanisms, this would add to the credibility of the conclusions.

A perspective study is based on observed events occurring in real time or close to the time of occurrence.[27] A retrospective study is based on observations made well after the events occurred. The Florida field study is clearly a retrospective, observational study, because the infection event (transmission from one infected tree to another) occurs months or even years prior to discovery of infected trees.

The Florida field study was also a census study as stated in the 2001 published article on the field study. A census study surveys 100% of the population within a fixed time period and within a designated area. During the survey time period, the population, namely the citrus trees within the sites, should not be changing. The population should not be affected by outside effects. Of course, no census study is perfect, but the flaws of a census approach should be identified as well as possible.

Confined and Unconfined Sites

An experimental study may be conducted in a containment greenhouse or an area where the escape of a pathogen would not cause any harm. Studies of pathogens which survive in semi-tropical or tropical conditions could be performed in USDA facilities in Beltsville, MD, in a temperate climate environment, to completely eliminate the risk of escaped pathogens.

Observational study sites may be open or confined areas. If the site is open, then there are opportunities for movement of pathogens to enter and leave the site. The Florida field study was conducted in unconfined sites within residential areas. This is discussed in Appendix A, with maps showing each site.

The various means by which citrus canker bacteria can move in and out of a site are shown in Figure 6.1. Three potential means by which citrus canker infections can increase in a site are: (1) Natural transmission of existing source trees within the site by windblown rain, (2) Natural transmission from infected trees outside the site and (3) The planting of contaminated nursery plants. In the case of nursery plants, the source of the citrus canker could be a hundred miles away.

Further, it is noted that one of the sites in the study (Site D1) was located one mile from the Broward county line, so contaminated nursery plants could easily be purchased, and planted in the study site during the study. For the other sites, citrus trees were sold everywhere in Miami-Dade County before 1995, and limited quarantine areas in Miami-Dade County were in place in 1996. The location of sites are shown is discussed in Appendix A.

.

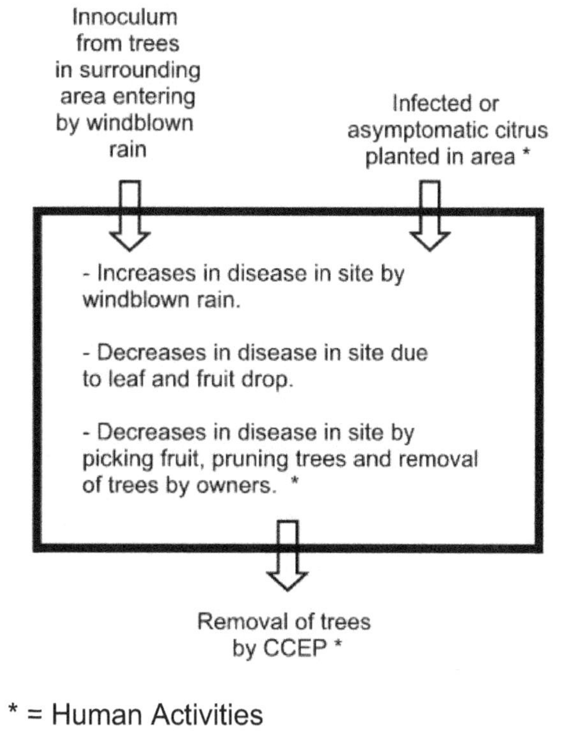

Figure 6.1: Site area with potential entry and exit of disease

As reviewed in Chapter 3, published articles suggested many other means by which canker bacteria could be transported with residential neighborhoods. However, these means of movement lacked scientific evidence and are considered just FDACS public relations misinformation. However, if one were to believe that citrus canker could be transported by contact with clothing, cars, lawn mowers, sanitation trucks, birds, insects and contaminated fruit, then the pathogens would move freely across the site borders, making it impossible to identify original and secondary incidences of the canker.

Attributes of the Population or Sample

The attributes are simply the essential properties obtained from a study. In a survey, there is normally a well defined observation window, where data collection occurs. Attributes may be fixed during the study, such as the location of trees and cultivars, or they may be changing during the study. The most essential attribute is whether a tree is categorized as infected or uninfected. This is based on the presence of lesions on a tree. In Chapter 3, the difficulties of definitive identifications of canker based solely on a visual examination in the field were discussed. The size of active lesions can be very small (1-2 mm) and difficult to see initially, as they may be high in the tree, sparsely distributed among new flushes. Older lesions tend to lose the yellow and brown halos and appear more like citrus scab, a fungal disease, as discussed in Chapter 3.

Residential backyards present additional challenges as there is a large variety of species and varieties which can be grown. There is no uniformity of the care given to citrus and infected fruit or leaves may be removed by pruning. Other similar diseases present such as scab and citrus bacterial spot may be present as discussed in Chapter 3.

Other information collected on both infected and healthy trees included citrus cultivar, tree age and location. The Department indicated that the locations of both infected and healthy trees were determined using GPS meters. Additional specifics of data collection in the field study are reviewed in Appendix A.

Wounds created by the citrus leafminer (CLM) allow the bacteria easier access to the mesophyll layers of citrus leaves. Unlike citrus canker, the presence of citrus leafminer is unmistakable. The symptoms are curled leaves with serpentine mines. At the time of the study, Dr. Gottwald and others had published articles indicating citrus trees with CLM infestations were much more prone to infection and seemed to support the case being made by him, for greater eradication radii. Yet, this information was not collected during the field study, which seemed odd, given the very unmistakable signs. A possible explanation is given in Chapter 7, Section 13, after a review of the epidemiology study.

5. Disease Triangle

Essential components of a plant disease are the necessary conditions which must be present for the disease to develop. As shown in Figure 6.2, the disease triangle is typically drawn to show the interactions of three critical elements, namely the pathogen, host and environment.

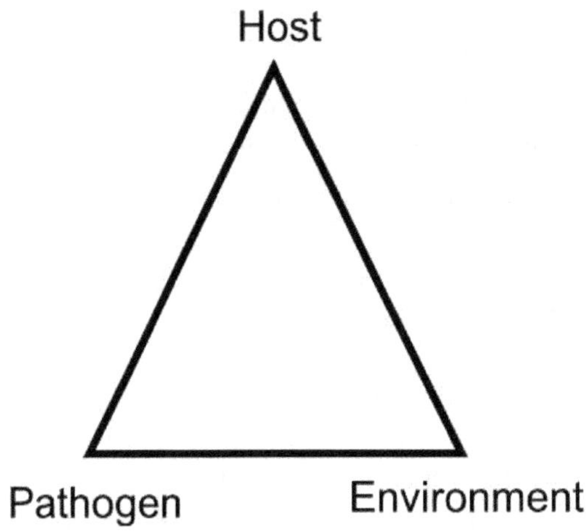

Figure 6.2: Disease Triangle

The infection process was described in Chapter 3 in terms of the pathogen and environment. The bacteria are exuded from the lesion when it comes in contact with water. The release, transport, deposition and entry can occur due to rain storms or irrigation systems. Multiplication of the pathogen is only possible if the bacteria come in contact with the citrus plant and enter the mesophyll layer.

A likely fourth critical element is human activities. Human activities may responsible for the introduction of the citrus canker disease into groves, nurseries, commercial outlets and residential.

Human activities include the purchase diseased citrus trees from commercial outlets by residents. At the time of purchase, these trees may have no lesions or very minute lesions. Nurseries and commercial outlets are inspected but still may have diseased trees in the asymptomatic stage. At nurseries and outlets, there is routine pruning of leaves to improve the appearance of the citrus plants and this can further delay the identification of citrus canker.

The human activities can add to the "elusiveness" of citrus canker, particularly in residential areas. Each owner may take care of their trees differently. Owners and lawn maintenance crews may trim off any visible signs of diseases, although the foliage may contain lesions in the upper branches.

In addition to pruning, environmental conditions (cold weather, high winds) as experienced in both Florida and Argentina during the winter months, can result in defoliation. Fruit and leaves with citrus canker lesions are more likely to drop.

6. Temporal Analysis

Temporal analysis is simply the evaluation of changes in the observed disease population with time. Since the disease populations changes in both in both time and space, temporal analysis is the observed behavior from one perspective.

There are different measures to assess the disease progress and intensity or concentration of diseases. If there is a survey of discovered infected trees in an area, then one could identify the number of lesions on a plant or the number of infected plants. Of course, this count data would not be meaningful in absolute terms. If the leaves of the plant are counted, and the number of lesions per leaves are identified, this is termed the severity of the disease, with a range from zero to one. Similarly, the number of infected trees per unit of area, termed disease intensity, with range from zero to one. A plot of disease intensity or severity, both denoted by variable y, with time is referred to as a disease growth progress curve.

It may be assumed that if $y = 1$, this would represent 100% of the trees are infected. However, if there is an active eradication program accompanying the survey, this can also an area in which by the end of the study, there are no infected trees. This point was made during Canker War I, where a plant pathologist stated that the map showing the spread of citrus canker, was in fact, where the Department knew canker no longer existed, as it had been eradicated.

The disease growth progress curve may be fitted to analytical equations for purposes of comparison of growth rates and extension into the future. Common analytical expressions used in curve fitting are exponential, monomolecular, logistic, Gompertz and Richards model.

In an article published in 1981, a study was conducted of the disease progress of citrus canker in commercial groves in Argentina.[5] The data were fitted to a Gompertz model as shown in Figure 6.3. Note, in the original article, the intensity was transformed to $y = -ln(-ln(y))$, to identify model parameters based on a straight-line fit of the data. The top curve, with $k = 0.18$, is a fit for the navel oranges while the bottom curve, with $k = 0.10$, is a fit for the other cultivars. It is also noted that the graph as shown in the published article does not extend the curve out beyond about 20 months.

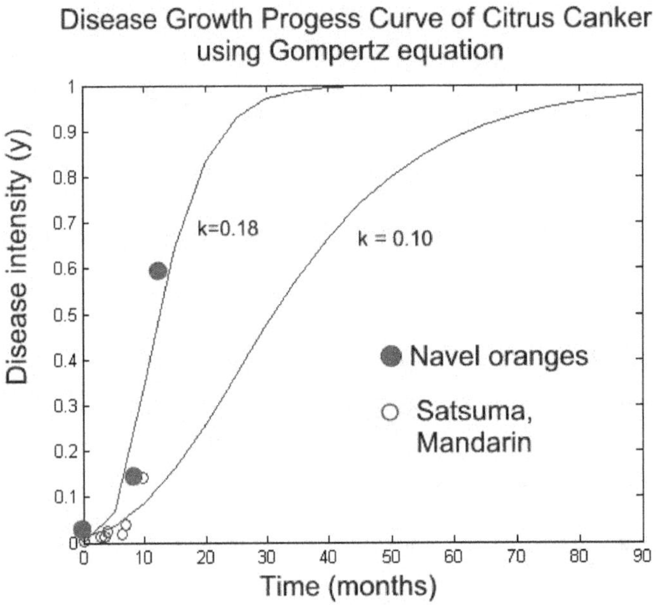

Figure 6.3: Disease Progress Curve

The fit of lower curve ($k = 0.10$), based on the Satsuma and Mandarin varieties of oranges, appears to be poorly supported by the data. It also appears the data could have been fitted to an exponential equation. The analysis is very dependent on a single data point occurring after 10 months. Data to confirm the model fit would require a very long experiment and perhaps infeasible as a grower would leave infected trees in place for as long as seven years. Variations in weather, normal leaf drop and pruning also complicate long duration experiments.

The intensity of disease can actually decline for significant periods of time. This is shown in Figure 8.1E, page 162, of "Introduction to Plant Epidemiology", by Drs. Campbell and Madden.

The disease showing a maximum and then a decline is an "early leaf spot" in peanut plants. The notorious Dutch elm disease also showed a decline in disease incidences.

Another case of declining disease incidences occurred in a study of canker dissemination in an Argentina nursery, as documented in Gottwald and Timmer's article, "The efficacy of windbreaks in reducing the spread of citrus canker by Xanthomonas campestris pv. citri, Trop. Agric., 1995. The cold weather cause leaves to drop, and the plants with disease symptoms were nearly gone after approximately 245 days. In other plots, the disease intensity continued to increase.

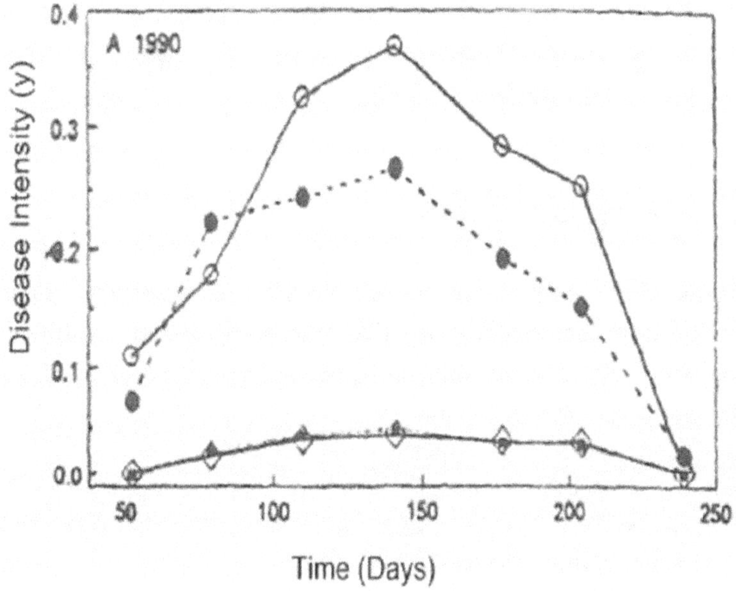

Figure 6.4: Citrus Canker Disease Progress Curve

(Ref: Gottwald and Timmer, The efficacy of windbreaks in reducing the spread of citrus canker caused by *Xanthomonas campestris* pv. *citri*, Trop. Agric. (Trinidad), July 1995)

The analytical or empirical equation approach represents a starting point in temporal analysis. Temporal analysis has been extended into the modeling of disease behavior using ordinary differential equations. The differential equation approach is an important improvement in bridging the gap between empirical approaches and the biological processes at play, but requires additional parameters. Where data are insufficient or unreliable to fully parameterize the model, this approach, nevertheless, may be beneficial in providing insight into the inter-relationships among factors involved in disease growth (Campbell[4], page 227).

There is considerable research in the area of theoretical modeling of plant diseases. With sophisticated computer software, numerical models are likely relatively easy to create. The success of these efforts in reliably predicting the temporal changes is likely to depend on the complexity of the overall system and the ability to identify the essential parameters.

There are certain difficulties in developing disease progress curves for citrus canker. In this book, the delay in time before identifying canker is termed the "observation lag" and may be more than a year after the first infection that citrus canker is definitively identified. Since there are many factors influencing the time lag, particularly in a residential epidemic, the disease growth curve based on new discoveries is likely be a distorted view of the actual growth of intensity with time. If there is an active eradication program concurrent with the study, then the actual disease intensity may be declining.

However, studies have been conducted to identify the overall disease growth progress curve for both groves and nursery settings where at least inspections can be done without concern for access to the property.

7. Spatial Analysis Concepts

Complex interrelations may limit the discovery of the full disease behavior in both time and space dimensions. Some diseases can be quantified in terms of changes in time, but not in space. Spatial analysis theory and applications lagged behind temporal analysis, as stated by Jeger, "It is not surprising that the overwhelming emphasis of the plant-disease literature has been on temporal aspects, as perusal of several important sources and textbooks will testify ..." .[18] This sentence ends with 10 references spanning 24 years. However, numerous advances in spatial analysis were well explained with the publication of "The Study of Plant Disease Epidemics" by Madden, Hughes and van den Bosch in 2007.[19]

An experiment to identify the spatial dissemination of a pathogen requires an inoculum source and an abundance of host plants which are disease free. If distances of transmission are to be determined, then experiment needs sufficient control to exclude any possible entry or dissemination outside the experimental space.

Observational Studies

In an observational study, it is necessary to identify the focal or source trees if inter-tree distances of dissemination are to be obtained. Depending on many factors, identification of these trees may be more difficult or possibly or impossible. Sometime nature helps in this process. For example, some crops die in the winter and are replanted in the spring. The disease survives the winter, and new sources of the disease are seen in crops during the spring and summer months. However, this disease cycle would not be applicable to citrus canker.

In Madden's 2007 textbook[19], page 174, it is stated, "Typically diseased plants do not move." Citrus canker should be considered an exception to this statement. Infected plants have been found in nurseries and later infected plants from these nurseries in groves, during Canker Wars I and II. Infected nursery plants were also discovered in 2005 during the 1900-ft policy phase of Canker War III. Citrus plants in nurseries are carefully inspected, and any plant determined to have citrus canker symptoms would be immediately destroyed. However, in the early stages of development, identifying symptoms can be problematic. Plants can not be assayed for disease because the disease in not systematic, meaning the bacteria do not enter the vascular system of the plant.

If citrus infected with canker escapes detection and are subsequently sold from nurseries, they act as point sources of canker in a commercial outlet, grove or residential environment in many locations throughout Florida. Within a residential area, one generally cannot survey all owners and find out when and where the trees were purchased. The supply chain for residential trees quickly becomes complicated by the movement of inventory between commercial outlets.

In the Florida field study, the Department stated the study sites were selected with recent discoveries of citrus canker. However, residential areas were never surveyed on a regular basis, so the absence of citrus canker in any study site or outside the sites can not be assumed. It is not known if these discoveries were the result of new introduction of the disease as a result of contaminated nursery plants. Focal or source trees were identified by a new technique ("Distance necessary to circumscribe method") as discussed in the technical review provided in the appendices.

Given the sources or focal trees have been identified in an observational study, it is essential that the pathway for introduction be identified, and strict measures are taken to ensure continued introductions do not occur during the observational period. Otherwise, any calculated distance from a focal tree to a subsequently infected tree may be invalid. If the pathway to introduction is the planting of citrus, then measures need to be taken to ensure this is not occurring in the study site.

Curve fitting

As with temporal analysis, disease intensity or severity, y, can be calculated as measures of disease concentrations at any location where there is citrus trees in the study area. The measure of disease intensity change with distance is referred to as a disease gradient. If the pathogen concentration has been determined at distances from a source, this is referred to as a pathogen gradient. The calculation of gradients likely will require the space to be discretized into subunits (or meshes) for changes in intensity to be calculated.

The observed changes in intensity with distance at a particular point in time, can be fitted to a mathematical model. Generally, locations are denoted as s, so the changes of intensity with respected to distance (disease gradient) is dy/ds. A common deterministic model of spatial intensity is the exponential model $y = a \cdot exp(-b \cdot s)$ which assumes $dy/ds = -b \cdot y$ and at $s =$

0, y = a. Examples of an exponential decline in disease intensity with distance is shown in Figure 6.5. The model is most applicable in a grove or nursery environment, where a continuous area of similar citrus trees are growing, as the deterministic model assumes the all areas of the space are available for disease development. In mathematical terms, one would state that a continuum for the process is presumed in models based on a differential equation or sets of equations.

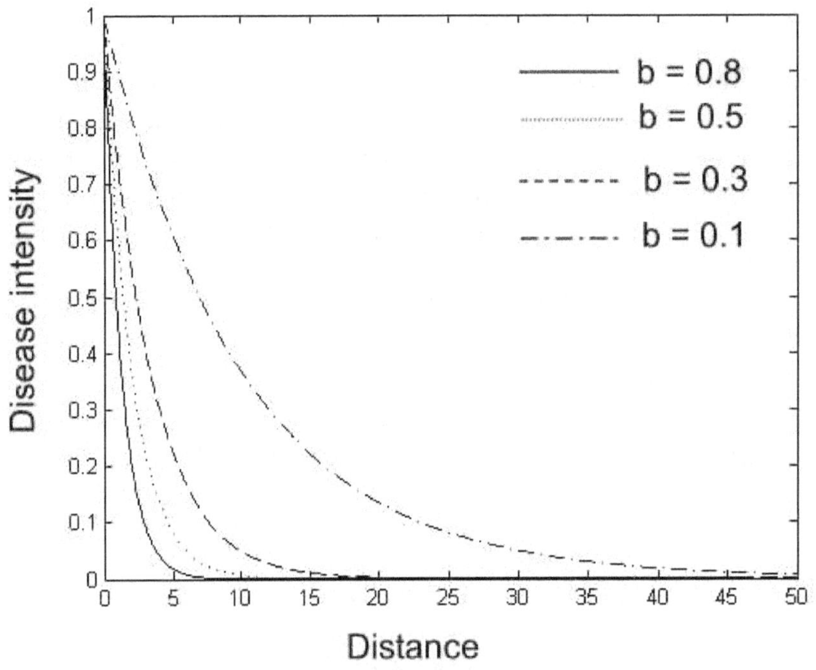

Figure 6.5: Exponential decline curves (a = 1 in all cases)

A graph of distance verses *ln(y)* would be a straight line in this case. Other models include probabilistic functions, resulting in a set of likely disease intensities with respected to distance. In all models, both the location variable, *s*, and the intensity variable, *y*, are continuous variables. This implies a continuum of locations where the disease can reside. A second aspect of the models, is there is no means of calculating the maximum dispersal distance unless one assumes that at a sufficiently low *dy/ds*, the disease is no longer effectively spreading. Otherwise, as *s* approaches infinity, *y* approaches 0.

More recent extensions of modeling efforts are to combine time and space component, either analytically or using simulation models. Further refinements incorporate additional disease components such as latency into a more comprehensive representation of the disease behavior.

Monte Carlo Simulation and Random Quadrat Sampling

The term Monte Carlo simulation refers to any analysis that uses repetitive generation of random number values in calculating results. This topic is reviewed since random number generation was used in the Florida field study as part of random quadrat sampling.

It is possible to create a spatial pattern that resembles a clustered pattern through Monte Carlo simulation. The process to create this pattern is referred to as a Neyman-Scott process.[24]

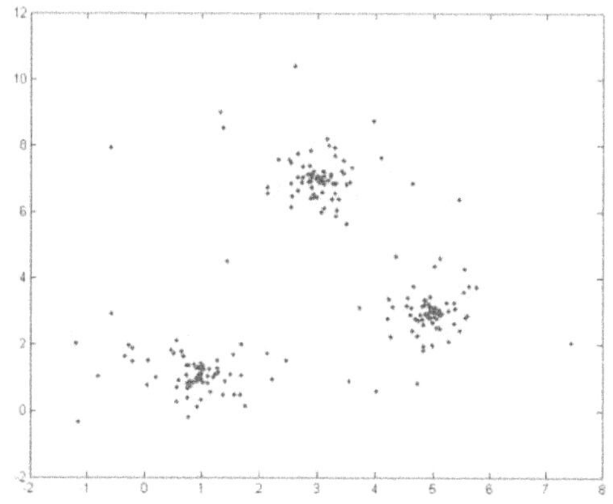

Figure 6.6: Computer generated spatial pattern

Both the initial infected trees (focal trees) and the subsequently infected trees from the focal trees are simulated by assuming probability distributions and related parameters. In Figure 6.6, three random points were created as source locations using a continuous uniform distribution. The distances from the disease sources were then calculated based on an exponential distribution.

In the Florida field study, the quadrat random sampling was used as discussed in the 2002 published article.[8] Statistics based on quadrat sampling include indexes of disease susceptibility, host density and disease severity. The steps to calculate these indexes are follows: (1) A random location (x,y) is generated within the study site which is the center of the square (quadrat) used for sampling (2) Data related to the trees such as height or cultivar/species within the quadrat are obtained (3) Statistics are calculated based on the quadrat sample data. In the field study, this procedure was repeated 500 times, so 500 index values are created. The Florida field study also used random quadrat analysis to create contour maps related to the presence of citrus canker.

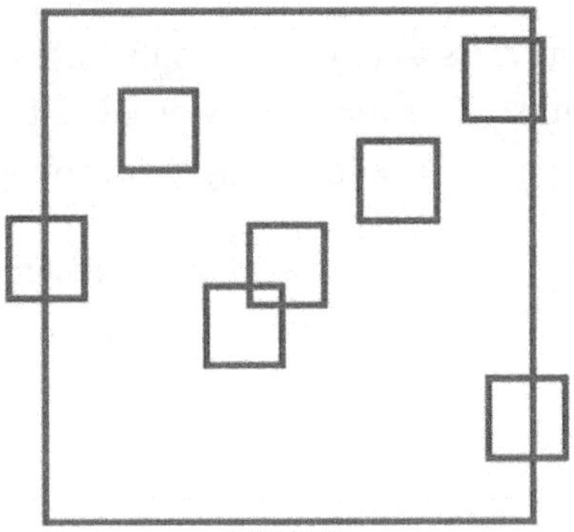

Figure 6.7: Random Quadrat Sampling

A thorough analyses of the sampling procedure and results as performed in the Florida field study are presented in Appendix E. It is noted that the typical application of random quadrat sampling occurs in survey areas where complete surveying is not possible. However, in the Florida field study, the application of quadrat sampling seemed odd as all residential yards were repeatedly surveyed.

Spatial Point Pattern and Geostatistical Analysis

Any set of points or points can be analyzed using numerous spatial methods including spatial point pattern and geostatistical analysis. The interpretation of the results should be done with caution, given full knowledge of the system, pathways of the disease and consistent with experimental studies. For observational studies of residential areas, the ongoing activities of residents need to be considered. Owners may pick infected fruit or trim branches with lesions. They may move or destroy citrus trees. They may also purchase and plant citrus trees from areas outside of the quarantine areas. These disruptive or complicating activities to spatial analysis are discussed in the critical review of the field study, as provided in the online appendices.

The Florida field study includes both spatial point pattern and geostatistical analyses. If a set of data consists of only the locations of infected trees as Cartesian coordinates, the arrangement of these points would be a spatial point pattern. In the case of the Florida field study, the citrus trees in residential areas are at irregular spacing, hence the locations of trees (healthy and infected) also constitute a spatial point pattern. The infected trees constitute a pattern within a pattern or a marked spatial point pattern.

If collected data consists of infected tree locations and a measurable property, then the data are amenable to geostatistical analysis or semi-variance analysis. The semi-variance analysis is commonly used in mining applications where the measurable property is the concentration of the mineral of interest. Normally, a variogram is calculated and values displayed on a graph with a best fit theoretical curve. Values at close distances should be strongly correlated, and those at longer distances the correlation should be weak or non-existent. The semi-variance analysis was used in support of long distance movement of citrus canker in the Florida field study as reviewed in Appendix G.

8. Selected Experimental Studies

There have been many experimental studies conducted on citrus canker, but only a few assessed the dispersal of inoculum beyond the immediate vicinity of the source due to windblown rain or similar simulated conditions. Three selected studies, as shown in Table 6.1, reported the dissemination of citrus canker bacteria in the range of 10 to 18 m (32.8 to 59.0 ft).

Table 6.1: Experimental Maximum Dispersal Distances

Study/ Author/ Source	Distance m (ft)	Pathogen or Disease Distance	Ref #
1990-91 Argentina Experiment Authors: Gottwald et al. 1995 Trop. Ag, Fig 2A	18 m (59.0 ft)	Disease Distance	11
USDA Experimental Set I Bock, Parker, Gottwald Published Jan 2005, Plant Disease	12 m (39.4 ft)	Pathogen Distance	1
USDA Experimental Set II Parker, Bock, Gottwald Published Dec 2005, Plant Disease	10 m (32.8 ft)	Pathogen Distance	21

The distances are identified as disease or pathogen distances. A disease distance is the distance from a source to a subsequently infected tree. A pathogen distance is the distance from a source to a location where viable cells of citrus canker bacteria are collected. In the above experiments, in the two USDA experiments, it is not known if the collected samples of cells were sufficiently concentrated in terms of CFU/ml to infect another citrus plant.

Argentina Experiment

The 1990 to 1991 Argentina experiment is documented in reference 11. The study was conducted to evaluate use of windbreaks and copper sprays to control canker spread. The maximum distance of 18 m was obtained from Figure 2A. The experiments consisted of 6 plots, with Duncan grapefruit (*Citrus paradisi*) in four plots and Villafrance lemon (*C. limon*) in two plots. The arrangement of these plots is consistent with a line source at the edge of a rectangular plot. The plot dimensions are not given, but the 100 healthy recipient citrus (grapefruit or lemon) are spaced approximately 0.30 meters apart, so the plot is 30 m long. The study plots consisted of

4 rows, with 0.76 m between rows. Plants were allowed to grow one season before the start of the experiment.

The maximum distance from the initial source to the furthest diseased plant was 59-ft (18 m) based on Figure 2A. However, since plants were not removed after infections were observed, the actual source could have been much closer than 59-ft. The experiment does demonstrate that for densely packed citrus plants, citrus canker can spread to a distance of 59-ft from a line of source trees in 250 days, which is the duration of the experiment.

The planting density is not provided, however it is estimated to be approximately 3300 citrus/acre, assuming an additional ½ spacing distance for an outside perimeter. This density is more consistent with a nursery density than a grove or a residential setting. In Chapter 3, the densities of nurseries, grove and residential settings were estimated to be 10,000 to 25,000 trees/acre, 129 trees/acre and 3.12 trees/acre, respectively.

USDA Experiment Sets I and II

The USDA experiments I and II are documented in Plant Disease, January 2005 and December 2005 issues, respectively.[1,21] Plant Disease is a journal published by the American Phytopathological Society (APS). Both articles are provided free to the public and not copyrightable.

Locations of the experiments are not provided, but likely conducted either in Florida or Texas, based on the office locations of the USDA researchers involved in the experiments. The experimental designs are similar, but the objectives are different. In the first set, a stated objective was to assess the dispersal gradient and distance to which *Xac* bacteria were dispersed in a simulated wind/rain event. In the second set of experiments, the stated purpose of the experiments was to compare techniques to sample *Xac* under conditions of windblown spray.

In both articles, wind was simulated using yard blowers. Water was sprayed from a nozzle into the air stream created by the blowers. The air stream passed through canker infected grapefruit tees. Downwind, the spray was collected in samplers consisting of a panel and funnel collection vessels. The first set of experiments, samplers were at distances of up to 12 m, while in the second set, samplers were at distances of up to 10 m. In both sets of experiments, bacteria populations were detected in the furthest samplers.

Given the experimental design, it can only be concluded that the experiments show dispersal of *Xac* up to a range of 10 to 12 m. The maximum possible distance of dispersal can not be concluded because there were no samplers beyond this range. Further, it can not be concluded that the recovered bacteria samples were sufficiently concentrated to cause infection, since no pathogenicity tests were done.

In the first set of experiments, the turbulence wind conditions were identified as wind passed over and through the grapefruit canopy. Upwind velocity was recorded at 19.5 m/s (43.6

mile/hour), while downwind velocity, recorded just 4 m from the source trees was 2.8 (6.3 miles/hour). The source plants were approximately 1.5 m (5 ft) tall, with canopies ca. 0.8 x 0.8 m, considerably smaller than productive commercial grove trees. The area effected by solid or porous obstructions to wind flow is termed wind shadow as discussed in a later section.

The first set of experiments also revealed other important information, including the decline in released bacteria concentrations with temperature and duration of the simulated rain event.

Other Experimental Studies

A series of experiments conducted in Réunion Island are documented in reference 23. Experiment were conducted to identify the release concentrations of *Xac* from simulated rain under windless conditions. The set up consisted of one infected citrus seedling in the center location surrounded by 24 host plants in an one square meter. The most significant result was that overwintered lesions had an order of magnitude lower internal *Xac* population than non-overwintered lesions. Bacteria populations were detected to corner host plants (~ 70 cm from source). This likely explains why citrus canker was easily eradicated in states north of Florida during Canker War I.

A series of experiments are conducted in Broward County, Florida in 2004 by USDA researchers and published in 2010.[2] The design was similar to the other USDA experiments, using an axial fan to simulated wind speeds of up to 19 m/s (42.5 miles/hours). Water was sprayed into the airstream from nozzles to simulate rain. The experiments varied wind speed from the fans, to assess the effects of bacteria dispersion. The maximum dispersal distance of *Xac* was 5 m, corresponding to the location of the furthest sampling vessels. It is unknown if the bacteria sampled was sufficiently concentrated to cause infection.

In addition to experiments specifically on citrus canker, in 1990, Drs. Gottwald and Graham published a study of the spatial analyses of citrus bacterial spot (CBS) in Florida nurseries.[9] The article included results from both experimental and observational studies. It is described as a uniquely different from Asian citrus canker, and notes that 23 million citrus trees with CBS were destroyed in the 1980's campaign. The prevalence of CBS in nurseries suggests that it may have been widely disseminated throughout Florida prior to eradication, but was more established in the southern area of Florida due to the more suitable climate conditions. Article has been posted on the online supporting documents website.

Wind Shadow

A wind shadow is an area of reduced wind velocity occurring upwind, downwind and above the canopy of a tree or any obstruction to wind flow. Wind shadow effects were observed in downwind measurements in experiment conducted by USDA scientists using a 1.5 m (5 ft) citrus tree. The wind velocity of importance in the transport of citrus canker would be downwind of the infected tree. As wind encounters impediments, the flow becomes increasingly chaotic, and eddies can form with counter-current flows, particularly in close proximity to the obstruction, resulting in areas of calm conditions.

Three factors influencing this area are the trees height, density and canopy size. Similar flow patterns occur with structures as shown in Figure 6.8. Large and small eddies can develop, as the wind encounters resistance. Canopy density is approximated by the canopy porosity, observed by the sunlight penetrating through the canopy. Citrus trees would likely be characteristic of a high density, as in general, very little sunlight penetrates the foliage.

The two approaches to assessing the effects of wind impediments or barriers are numerical simulation and experimentation. The work is motivated by a need to protect agriculture and soil from high winds. As a general rule, an area of reduced wind (50% or more), exists for a distance equal to 5 to 10 times the height of a tree. The distance to full wind velocity can be 20 to 30 times the height of a tree.[3, 28]

Wind shadows would likely have an effect on the dissemination of citrus canker in residential areas where houses and other structures (apartments, offices, schools) are present. The Florida field study was conducted in areas which contained both commercial and residential buildings. For an infection event to occur between two backyard citrus trees (one tree infected and the other healthy) on opposite sides of a street, assuming the wind is blowing in the direction of the host tree, the rain water droplets containing bacteria would have to pass through the infected tree and over the house on the first property, across the street, and pass over the second house, to finally land on a citrus host in the backyard of the second property.

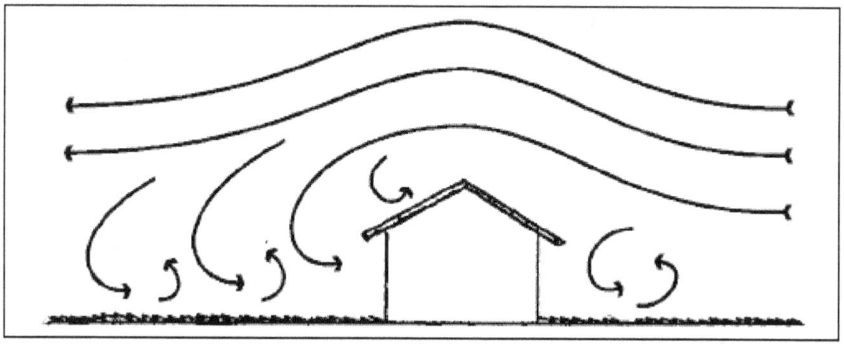

Figure 6.8: Wind shadow and zones of turbulence due to a structure [28]

It is noted that after considerable literature search, no published experiments could be located which demonstrates canker infected tree close to one side of a house can transmit canker through windblown rain to a citrus tree on the opposite side of a house. There are turbulent downdrafts or eddies on both the windward and leeward side of the structure, making the pathway difficult.

Ideal Experiment

The experimental studies, as discussed in this chapter, were not specifically directed at identifying the longest possible dispersal distance. However, distances from source to the point of collection vessels are provided. It is suggested the ideal experiment would entail a single infected plant or set of multiple infected plants to form a point or line source. This source would be surrounded by non-infected host plants located at various distances. The source plants are easily be created by inoculation, followed by an incubation period. The experimental site would be isolated from all other citrus.

After every rainstorm or synthetic storm using water sprayers/misters and blowers place behind the source plants, the healthy plants potentially contaminated with bacteria, would be removed and placed in a containment greenhouse to be observed. Each storm event could be short, as the highest concentration of inoculum released from lesions occurs within the first hour or less of a storms.[23] Over a period of several months, thousands of trials could be performed.

The reason why the contaminated plants need to be removed, is to remove all possibilities of secondary sources to be generated. Thus the experiment would only determine the distance from a known source of trees to other healthy trees.

An ideal experiment could provide the probability that a plant with severe wounds or citrus leaf miner infestation would be more likely become infected from a simulated rainstorm. Of course, there would need to be many experiments conducted with host trees at various distance, and variations in wind and rain conditions, to obtain sufficient statistics.

Florida is not lacking in resources to conduct experiments even with extremely high winds. Florida International University in Miami, has facilities to simulate hurricane force winds (> 100 mph). Further, if an experiment was conducted to reproduce canker dispersal in residential areas, barriers should be set up for the effect of houses, cars and other potential barriers to transport

To the knowledge of this author, no full experiment has ever been published in the literature at least not up to the submission of this book for publication in 2016. There are many ongoing projects, and in time, there may be some study published in the literature, which fits this definition. If so, links will be provided to the study on the online supporting documents website.

9. Selected Observational Studies

The focus of this book is the Florida field study as documented in two published articles, a Letter to the Editor in year 2001[13] and more extensive article in year 2002.[14] According to the authors, this is the only field study of citrus canker in residential environment. As of year 2016, it still appears to be the only study of citrus canker anywhere in the world, which considered disease dissemination within residential areas. The Florida field study is thoroughly reviewed in the appendices, available online.

In 1980, Stall et al. published an article entitled," Population Dynamics of Xanthomonas citri causing cancrosis of citrus in Argentina"[25] in the Proceedings of the Florida State Horticultural

Society. This was a brief research note which is generally credited with the scientific support of to the 125-ft rule. The authors describe collection of rainwater samples after showers occurring on 5 days: October 18, 1978, March 7, April 5, November 8 and December 13, 1979. The furthest documented distance of dissemination of viable bacteria was 8 meters, from the data provided. The authors state that cells of citrus canker were found in collected in rain water samples approximately 30 meters from the initially infected trees on December 13, 1979, but it was never determined if these cells were capable of causing infection. A review of the research is provided in a short note, provided on the website.

In 1984, Danos et al. published a study of the temporal and spatial spread of citrus canker in Argentina.[5] Repeated surveys of the study sites showed increasing incidences of citrus canker with time and distance, and these results were fitted to equations as functions of time and distance. The study does not present the dispersal of citrus canker in relation to weather events.

An analysis of the dissemination of citrus canker in a Florida grove was published by Gottwald et al. in 1992.[10] One of the more important conclusions, as given in the abstract of the article was, "In October 1990, the occurrence of Asiatic citrus canker in an orchard in south Florida was apparently related to spread of *Xanthomonas campestris* pv. *citri* from dooryard trees 230 m away in an adjacent property." The analyses as provided in the article is discussed in a short note identified as the "Highlands Observational Study."

Four more recent studies are noted below. Since all of these studies were published after the USDA and FDACS concluded that further eradication efforts were infeasible in Florida, the studies would not have a direct impact on eradication program. For this reason, the articles are not reviewed within this chapter.

(1) Brazilian study, (2007)[16]: The dissemination of citrus canker in selected groves in Brazil ("Brazilian study"), was analyzed by Gottwald et al. in a study published in 2007. The authors state that the scientific basis for the Brazilian eradication program to destroy all canker infected and healthy citrus in commercial groves to 30 m is based on the 1980 article by Stall. Further, the authors suggest that the pattern of dissemination in groves in Brazil in sufficiently similar to those observed in residential areas of Florida, resulting in similar distances needed to circumscribe subsequently infected trees. The result is surprising given the differences in host densities (commercial grove ~ 130 citrus/acre, residential ~ 3 citrus/acre). The article states that 326 maps of citrus plantings were examined, which indicates an extensive set of collected data. Citrus groves in Brazil are subdivided into blocks. For control purposes, this is very important, as if citrus canker incidences exceed 0.5%, all citrus within a block are destroyed. The size of the blocks are not given in the article, which is a critical information. Brazil does not conduct extensive surveys and healthy tree eradications within residential areas.

(2) Hurricane Studies, (2006 and 2007): [17, 15] Two articles were published on the dissemination of citrus canker by hurricanes in the development of a predictive model of canker spread following catastrophic weather events.

(3) Optimal Strategies Study, (2009):[22] A study on the optimal strategies for eradication of citrus canker was published, using used Monte Carlo simulation (Neyman-Scott process) as discussed in section 6 of this chapter. While this study is not strictly an observational study, the parameters used in the probability distributions relied on the Florida field study.

(4) Bayesian Analysis Study (2014):[20] In 2014 an article by Neri et al. was published to further analyzing the Florida field data using Bayesian analysis. Bayesian analysis is a statistical procedure which endeavors to estimate parameters of an underlying distribution based on the observed distribution. This article provided maps showing the locations of infected trees in the Florida field study, which is useful information in the review of the Florida field study as presented in Appendix A.

10. Concluding Remarks

The chapter emphasized terminology and concepts rather than detailed technical discussion. Studies in general can be classified as observational or experimental. Also studies can observe events or the results of events in the present (perspective) or past (retrospective). Study sites boundaries may be open or closed. Data may be collected by sampling or by complete and repeated surveys.

The study of diseases within residential areas has unique complications. Unlike diseases which are vectored by insects or carried long distances, citrus canker requires wet conditions for both release and entry into host plants. Houses act as windbreaks, impeding disseminations. Residential areas have large non-citrus areas. Unless experiments show that canker can travel by windblown rain over these areas, then it would be incorrect to calculate transmission distances from source to host tree, which transverse these areas.

A literature search was done of all experimental and observation studies on citrus canker and a related disease, citrus bacterial spot. Three experimental studies were reviewed in detail. From these studies, the maximum distance of dissemination is 59-ft was estimated. Links to these articles are available on the website.

The next chapter presents a summary of the Florida field study and related problems. However, the detailed technical review of this field study is provided in the appendices, and should be read next and then the reader should continue to Chapter 7.

Short Notes on Website

SN 5.1 Peer Review of the Field Study
SN 6.1 Argentine Observational Study (1978 to 1979)
SN 6.2 The 1900-ft Policy and the Published Articles by Dr. Gottwald et al.
SN 6.3 Comparison of Department Website Justification Statements
SN 6.4 1990 Highlands Observational study
SN 6.5 The Big 3 Articles on Citrus Canker

Part II:

Going Beyond the Basic Facts

7. Field Study Investigation Summary

The plural of anecdote is not data

Roger Brinner

Also on hand was Mark Fagan, the public information officer. He tried to explain, "This tree may not look sick to you, but in a few months, it will start losing its canopy and fruit won't mature any more," Fagan said sternly. "And the disease will spread another 1900 feet to other trees."

"That's bullshit," fumes Maria.

Anatomy of a Quarantine, Miami New Times, Kirk Nielsen, July 6, 2000.

---

1. The Detailed Investigation

The online appendices of this book present a thorough investigation of the 1998 citrus canker field study. This chapter summarizes the conclusions reached in these appendices. There were six separate analyses of the collected data. For each analysis, a brief summary of the analysis performed and the validity of results are provided in this chapter. A list of the appendices is provided at the end of this chapter.

The summaries on the analyses perform may be too brief for many readers. Hopefully the summaries will entice readers to examine the full review presented in the appendices. Often, authors relegate material of lesser importance to the appendices, so the flow of the narrative can continue without an overload of technical details. However, in this case, although the investigation is technical in nature, it is not of lesser importance. In fact, it is essential to this chapter.

Every effort was made to evaluate the study on a "bottoms up" approach. This involved examining the data collection and study sites, prior to a review of the methods and results used in the study. Ideally, a bottoms up approach would begin with examining the study plan, any changes to these plans, survey forms, GPS meters training, instructions to inspectors, their training prior to inspections, their background and all computer software used in the study. However, none of this information has been made public.

The primary article on the field study was published in Phytopathology by Gottwald et al in April 2002.[8] For convenience, this document is referred to as the "2002 article." An initial summary of results was published in 2001 as a Letter to the Editor in Phytopathology by Gottwald et al.[7] Other documents released by the Florida Department of Agriculture and Consumer Services as a result of legal action in November 2000 were also useful. These documents include reports from the Citrus Canker Risk Assessment Meeting in May 1999[3], the interim report of the field study, as received by the Department in October 1999, and copies of the viewgraphs of a presentation made by Dr. Gottwald in Broward Court[6] in November 2000. These documents are provided on the website.

The study's objectives were identified in Commissioner Crawford's press release in February 1998.[1] The Commissioner asked scientists to track the movement of the citrus canker disease in three specified sites for one year, and meet every month to discuss the results. The study was to coincide with the moratorium on healthy tree cutting in Miami-Dade County.

2. What the Researchers/Department Would Like Residents to Believe

The Department wanted residents to believe good science lead to the 1900-ft policy. A second message was that the Department was guided only by the best researchers. Finally, the 1900-ft radius was portrayed as distance less than the maximum distance citrus canker could spread during a storm, but was still sufficient to eradicate the disease. For both public relations and legal purposes, the Department wanted to make it clear that it was exercising restraint when it implemented the 1900-ft rule.

This is a short summary of findings as shown below was taken principally from the published 2002 article on the study.[8] However, other statements of findings, including Dr. Gottwald's testimony in November 2000 in Broward Court[6], and abstracts of presentations made at American Phytopathological Society meetings were also considered.

- **Distance Necessary to Capture (DNC) Method:** The "distances of spread" helped regulators in an adjustment of the eradication radii. An average "distance of spread" is not provided in the 2002 article, but there is a general statement that 1900-ft is within the range of values of distances necessary to circumscribe in the first four periods of the 2002 published article.

- **Weather Analysis:** Citrus canker lesions are at the maximum visibility, on the average, 107 days following a rain storm with wind.

- **Inter-point distance analysis (IPD):** When all possible distances are considered, the maximum inter-tree distances ranges from 914 m (2,999 ft) to 4754 m (15,597 ft), for all study sites.

- **Quadrat Analysis and Mapping:** Using an index from random quadrat sampling, the development of citrus canker as it develops was shown in a "real" time sequence. Contour maps show development of new foci, followed by infilling.

- **Spatial point pattern analyses**: These analysis provided further evidence of long distances of spread are possible.
- **Semivariance analyses:** Changes in the range attribute (RSTD) indicate rapid increases in disease development in both the spatial and temporal scales.

The Department's short summary of the epidemiology research changed as the eradication program progressed as documented in Short Note 6.3.

Peer Review Science

On numerous occasions, the Department has suggested, that residents should consider the 1900-ft rule as proper because it was based on peer reviewed science. The two articles on the field study, one published in January 2001 as a Letter to the Editor[7], and the second article published in April 2002[8], were peer review by the American Phytopathological Society (APS), which is a well respected international society focused on new research in all aspects of plant diseases. Their website is www.apsnet.org. It is noted that the April 2002 article provided details on the field study including methodology, data collection, results and conclusions as required by the APS.

Neither of the two articles on the field study were published prior to the implementation of the 1900-ft rule. The only document available for review was the October 13, 1999 interim report, which the Department has not publically distributed, except as evidence in the Broward Court in November 2000.

The issue of peer review only seems relevant because the Fourth District Court of Appeals accepted the idea of the Department that the 1900-ft policy was, in some manner, a scientific result based on the publication of the April 2002 article. The process used by the APS to select articles suitable for publication through the peer review, is discussed in Short Note 5.3. Peer review does not mean the reviewers or Editor-in-Chief accept the results, but that the article satisfies certain criteria for submission. In fact, at times publication allows others to publish a rebuttal to the articles, identifying the problems with studies.

3. Real and Synthetic Chronology

The real chronology is the record of when a tree was discovered with citrus canker. Each discovered infected tree has a second date associated with it, called the initial infection dates or IID according to the 2002 published article.[8] These are calculated dates, obtained from the discovery date minus the age of the oldest lesion on the infected tree. These dates are used to calculate the "synthetic chronology" of events. The raw data, which would include the discovered date, age of the oldest lesion and initial infection date for each infected tree has never been released by the USDA/ARS.

In a presentation in Broward Court in year 2000, Dr. Gottwald presented a selection of lesion ages ranging from 4 to 10 months.[6] Since the lesion age is shows considerable variations, the synthetic chronology may have little in common with the "as discovered" or the real chronology.

All results are presented in the published paper according to synthetic chronology. This makes it impossible to know the state of the site was prior to the study or what was discovered as infected trees during the study. For example, it is unknown the number of discovered infected trees in February 1998, when Commission Crawford first announced the study, nor in December 1998, when Dr. Gottwald presumably made a presentation of results at the USDA office in Orlando. The number of infected trees at the end of the study is known, but the date on which surveys ended is unknown.

The use of the synthetic chronology appears to be unique to the Florida field study, at least in the published technical literature. After a considerable literature search, no other article could be located, which used this approach.

4. Inspections and Study Sites, Appendix A

What was done

The details of the surveys of residential properties as described in the primary article on the study[8] in conjunction with other public documents are presented in this appendix. There were three sites in Miami-Dade County and two sites in Broward County in the 2002 published article.

The study required multiple entry onto properties to identify healthy and infected citrus trees. The inspections identified the citrus trees and whether they were infected with citrus canker or healthy. In addition, the age of the oldest lesion on the tree in terms of days, had to be identified, as this is necessary information for the "Distance Necessary to Circumscribe (DNC) Procedure." Parcels would be inspected approximately 60 days.[8] Earlier presentations of the study indicate that monthly inspections of all properties were made.[3] It is presumed that on subsequent visits to a lot, the same information would be recorded.

The published article[8] provides no information on inspections of residential lots outside the boundaries of the study sites. FDACS website summary of the epidemiology study indicates the area outside of the sites were free of infected trees at the onset of the study.

It is noted the Department held a series of public hearings in 2001, as required as part of passage of new eradication rules. None of the scientists involved in the Florida field study (Drs. Gottwald, Graham, Sun, Riley, Ferrandino, and Hughes) were in attendance at any of these hearings.

Did the surveys provide reliable data?

The same problems which plagued the eradication program, likely existed during the field study. These problems include difficulties in accessing backyards, identification of citrus canker particularly in the early stages of development and full inspection of back yard areas, particularly in the summer which in well known for afternoon downpours. Visits to all sites indicated Site D1 (Carol City) was most likely to have inspection problems, as it is an economically depressed area. The most common deterrents against break-ins were chain link fences and guard dogs.

The Commissioner's press release in February 1998 was very short and lacking critical details, such as the relationship between the field study and protocol changes. The press release stated nothing about the effectiveness of the current 125-ft policy nor a need to enlarge the eradication circles. According to the press release, the choice of the study sites would be subject to the approval of the USDA. Beyond this, the press release did not specifically state which organizations would be involved in the study.

As previously stated, the principal investigator of the study, Dr. Gottwald, testified in Broward Court in November 2000, that the 1900-ft policy was not agreed upon based on the results of any report.[6] In year 2000, public relations officials with the Department, were explaining the 1900-ft rule was based on a large field study with 19,000 citrus trees, led by Dr. Gottwald.

Basic information on the field study, including time of inspections, study site locations, survey sheet entries and how conflicting data were reconciled, was either missing, inconsistent or contradictory. For example, it is unknown when inspections began and ended in each of the sites. The Broward sites were added some time after the start of the inspections in Miami-Dade County, but it is unknown when this occurred. The only date known is, February 26, 1998, when Commissioner Crawford announced the field study.

Without this information on the start and end dates, it is unknown how many visits to yards were made in each site. Likely, there would be times when a visit would be incomplete, either due to access problems or weather events. Return visits to yards were never discussed. How could inspectors know they had been inspecting the same trees? The inaccuracy of the GPS meters (+/- 25 ft) was well documented, and likely inadequate for determining tree locations with a back yard. It is assumed that the determination of the oldest lesion age was done at the time of inspection or shortly afterwards.

If the inspectors did not have the prior inspection sheets, they may have inspected the lesion ages on the same tree twice. But, if there were blemishes on the foliage, it may have simply dropped off. Or the oldest lesions may have been on the very top leaves, as new leaves are the most susceptible to canker, but the most difficult to see. Owners may also simply trim off infected leaves. Obviously, if they thought the Department was going to destroy their tree if they found evidence of canker, so owners would be motivated to "clean up" the tree.

The conflicting information regarding the study site locations is reviewed in detail in Appendix A. It was necessary to examine documents from presentations and published articles, in this reviewed. Many discrepancies in the site boundaries are noted.

5. Distance Necessary to Circumscribe (DNC) Method, Appendices B and B1

What was done

This is a short summary of the DNC method. It is thoroughly described with examples in Appendix B. It was first described in the 2002 article.[8] No prior application is cited in the references. There appears no similar application of this method in Phytopathology, at least to year 2016. Selected results have been provided in several documents including a Letter to the Editor of Phytopathology[7] in 2001 as discussed in Appendix B. without a formal description.

The DNC method is a theoretical means of determining a set of distances, which the authors refer to as "distances of spread" or "distances necessary to circumscribe or capture." Each infected tree is associated with another infected tree, deemed the tree responsible for the infection. The DNC method is consistent with a "parent-offspring" model, where every offspring is matched with a parent. At some point, the offspring becomes a parent, capable of infecting more trees. For this review, parent and offspring trees are referred to as prior infected (PI) and newly infected (NI) trees, respectively.

The field study article[8] refers to PI and NI trees as focal or alpha trees, and secondary infected trees, as given in the article in tables 1 to 5, columns 4 and 5. The steps, as given below, relate only to trees infected to citrus canker.

1. IID Calculation Step: The IID of each tree is calculated as IID = discovery date - age of the oldest lesion.

2. Time Period Parsing Step: Each infected tree is assigned time period based on its IID as a new infected (NI) tree. This creates twenty-four 30-day time periods, each with a discrete number of infected trees.

3. Near Neighbor Association Step: For a particular time period, all trees in the prior time periods are PI trees. Any infected tree with an IID less than the start date of the first period is a PI tree or potential parent for infected trees in first time period. The method chooses the nearest PI tree to each of the NI trees as the responsible PI for the infection and calculates the inter-tree distance for each NI trees. The number of distances must equal the number of NI trees within the time period.

According to this step, two infected trees may be next to each other, but if they are in the same 30 day holding period, one tree can not be responsible for passing the disease to the other tree. There are other cases shown with longer holding periods (60, 90, 120 days), with different rules, as explained in more detail in Appendix B.

4. **Statistical analyses:** The prior step results in a set of distances for each site, equal to the number of discovered infected trees. The method of calculation of the 95th longest distance is not discussed in the 2002 published article[8], but is assumed to be based on percentile estimation.

From examination of five other presentations of results prior to 2002, it was concluded in Appendix B1 that the DNC method was not applied in a consistent manner, as some of the distances would exceed the maximum distances within the study sites. It was concluded that these distances were calculated by associating infected trees from within the study site with infected trees far beyond the study site boundaries. In the next time period, trees outside the site would not be used further in distance calculation. Thus, this alternative means of associating trees is termed "Use them and lose them" method in Appendix B1.

Are the DNC method and results valid?

The most cited results of the DNC method, are the "distances necessary to circumscribe" with a 95% probability with the 30-day temporal windows, as presented in the primary published article. These distances ranges from 0 to 11,608 ft, when based on the results from all study sites and all time periods.

The central problem with the DNC method is that it requires accurate identification the age of all lesions on a citrus tree, to determine the age of the oldest lesion. An error as small as one day in the oldest lesion age can result in the incorrect assignment of a tree to a temporal period. It is shown that an error of 10 days in estimation of oldest lesion age will result in an 18% probability that a tree will be assigned to the incorrect temporal period. The size of the lesion in these 10 days would increase by 0.30 mm, assuming an approximate 0.10 mm/month as given by Schubert, et al, in the 2001 article. All lesions can not be identified in a citrus tree, and the oldest ones are likely to be high in the canopy. Uniform expansion does not occur, as over time, the yellow and brown halos can diminish or even disappear, and citrus canker can resemble citrus scab, a fungal disease.[12] In Argentina, most misidentification occurs when only older lesions are present.[13]

Further problems with the distance calculations are discussed in Appendix B1. In this appendix, unusual long distances from 1999 to 2001 presentations were investigated. These distance results would not fit within the sites. It was further discovered that the April 2002 published article also used infected trees located outside of the study sites. This was evident in a 2014 published article by Neri et al[10], which provided maps of infected trees used in the study. In study sites D1 and D2, it is clear that the field study did not limit the observations to the square mile sections, defined by sector-township-range (STR) as stated in the published article.

These are serious violations of the study's controls. Since the distance from any infected tree could be related to another tree, whether it was inside the site, or beyond the site, then any infected tree within South Florida could be related to the source trees within the site. Without controls, the larger distances shown in the tables are arbitrary, and are dependent only on the discretion of researchers on how far beyond the site boundaries are acceptable.

Under these conditions, none of the results in the April 2002[8] and January 2001[7] articles should be reliable estimates of inter-tree disease travel distances.

6. Weather Analyses, Appendix C

What was done

Correlation analysis as presented in the 2002 article showed a correlation relationship with two daily time series, a weather index and canker incidences. For Sites D1, D2 and D3, a maximum correlation coefficient (r^2) of 0.988 results at an offset of approximately 107 days, which is termed as the "maximum visibility" of canker lesions. Additional time offsets of each study site were provided in the October 13, 1999 interim report as follow:

Table 7.1 Cross-Correlation Analysis (from October 13, 1999 Interim Report)

	Disease x Precipitation X 100		Disease vs. Precip. X 100 X Wind Gust	
Study Site	Offset (days)	Corr. r^2	Offset (days)	Corr. r^2
1	59	0.988	101	0.987
2	55	0.983	111	0.982
3	198	0.959	198	0.962
Broward	-8	0.995	13	0.986
Total (All Sites)	39	0.991	79	0.986

All correlation coefficients in this range (0.959 to 0.991) would generally be considered excellent, and indicate a high level of certainty that a linear relationship exists between the two variables.

For each site, the coefficient is computed from two cumulatively summed time series. The first series is the weather index, based on the product of rain and wind velocity. The second series is the citrus canker incidences occurring in the sites over the study period of 540 days. The incidence curves used the same as in Tables 1 to 5, which are a direct result of the parsing of infected trees into finite time periods based on their IID's (Step 2 of the DNC method).

Are the weather analyses valid?

When variables which are obviously unrelated show high correlations, this can be by design or by accident. The most common textbook example is of the "by accident" type. For instance, there may be a positive correlation between the number of Americans living past 65 years old, verses the sale of Coca-Cola when compared as time series. Both variables are likely increasing with time, likely influenced by an increasing population (a lurking variable) and no causality should be associated between these variables.

However, in this case, it is clearly a case of spurious correlation by design. A statistical trick was used to show excellent correlation between incidences of citrus canker and weather events. The two time series (disease incidences and weather index) are the result of cumulative summation of their underlying series. After the cumulative summation, the series will inevitably show high correlation value even if the underlying series are composed two random independent values. An example of this statistical trick is shown in the Appendix C.

In order to confirm this conclusion, the correlation analyses was duplicated for Site D1. The results were nearly the same for the cumulatively summed series, however when the series was not cumulatively summed, the maximum positive correlation coefficients is 0.1219 at a time offset of 441 days. This supports the conclusion that the correlation between weather events and disease incidences based on IID time scale is very weak or non-existent.

Also, the concept of a single "maximum visibility time" appears contrary to the biological process. It would imply that the canker becomes less visible after this point in time. The lesion becomes more apparent and distinguishable from other foliar diseases with time.

Finally, the disease incidences are not on a "as discovered" basis, but on the initial incident date basis. These dates takes into consideration the time of latency. As such, if these dates were correct, then the offset time of 107 days has nothing to do with the time to discovery of citrus canker or the time to reach maximum visibility.

This is not to suggest that a time lag does not occur between when there is a rainstorm, and the subsequent incidences of citrus canker can be observed. For many reasons, there may be an "observation lag" ranging from a few months to a number of years, depending on many factors. Obviously, developing relationships are more difficult in residential areas, particularly if inspections are infrequent.

The weather correlation analysis was used in the 2002 article to suggest more frequent inspections could be used with the 125-ft policy would be ineffective in eradicating citrus canker. This was discussed in the Technical Task Force Meetings in 1999. The weather analysis was presented by Dr. Gottwald at least three times: (1) October 13, 1999 interim report, (2) Broward Court presentation in November 2000, and (3) the 2002 published article. It was cited in the Fourth Court of Appeals opinion as evidence of scientific study.

7. Inter-Point Distance Analysis (IPD), Appendix D/D1

What was done

The IPD method calculates a large set of inter-tree distances considering any tree could be the source tree for the other newly infected trees as shown in Figure 7.1. The IPD method was applied to all study sites. The published article[8] presents a frequency distribution graph along with the peak and maximum distance values. The article indicates these values represent "overestimates" of the distance of spread.

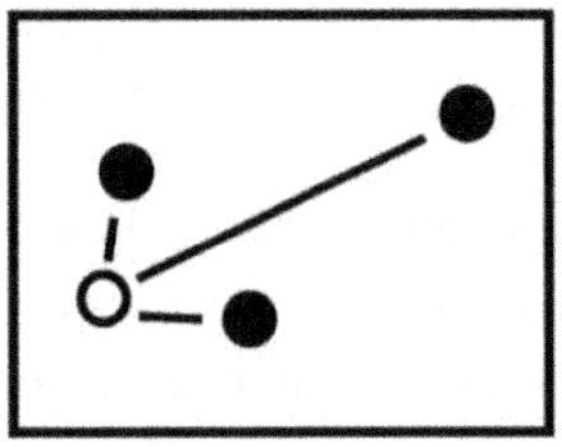

Figure 7.1 Inter-Point Distance Procedure

Is the IPD analysis valid?

The method calculates thousands of hypothetical distances. The calculated maximum distances for each site range from 0.6 miles to approximately 3 miles. The frequency distribution is often referred to as a histogram, which is useful in estimating probabilities of events. The frequency approach to estimating probabilities under the presumption each distance value is possible and equally probable. For example, a calculated distance of 50-ft would have the same probability of occurrence as one 5000-ft away. This leads to absurd conclusions on the possible range of dissemination.

However. the authors never refer to the frequency diagram as a histogram, and typical descriptive statistics of a distribution such as the mean and median, were not presented. As presented, the frequency diagrams are seemingly correct results, but the interpretation that these distances relate to dissemination distances or "distances of spread" by storm events is improper.

The article states that the values presented in the frequency diagram provides an overestimation of distances. For a widely dispersed disease, the peak and maximum distances are functions of the area of the site and geometry, and are unrelated to any inter-tree travel. If a similar study consider all infected trees in Florida, the maximum frequency values would have been on the order of hundreds of miles.

Further confusion in the results is identified in Appendix D1, where the number of distances in the article is nearly twice the number of distances which should result, if the procedure as stated had been followed. It is likely that the authors are not calculating distances between source and newly infected trees, but calculating distances between all infected trees as discovered in the sites.

Physically, these maximum frequency values of up to 3 miles distance as travel distances are impossible. Assuming the lots are approximately ¼ acre squares (104 x 104 ft), for rain water to go 3 miles, with citrus canker bacteria, it would have to travel across 150 lots to reach an infected tree. The rain drop would have to first pass through the canopy of an infected tree, and then go over fences, canal, lakes, roads and perhaps 150 houses, to infect another tree.

8. Quadrat Sampling and Related Analysis, Appendix E

What was done

A quadrat is an areal sample. In the field study, it was a square, 0.5 km on a side for a total area of 0.25 km^2 (61 acres). As described in the 2002 published article, the field study used random quadrat sampling, where the Cartesian coordinates of the quadrat centers were chosen using a random number generator, limited to the bounds of the study site.[8]

Random sampling is used to provide an unbiased sample of a larger area.[9] Through random sampling, a property or attribute of the disease can be assessed, such as the number of infected plants per unit area, and trends or other patterns may be inferred.

In the published article, the random quadrat sampling generates statistics for each site which are used in two analyses: (1) A linear regression analysis of variables with an indexes of disease severity and normalized disease incidences, and (2) Contour maps showing citrus canker development in a time sequence. Regression results are shown in Figures 6A to 6J and contour mapping in Figure 7 in the 2002 article.

Linear regression forms a relation between an independent variable, x, and a dependent variable, y. The authors describe a fairly complex procedure of creating various index variables in the primary article, which is reviewed in Appendix E.

The citrus canker development in time as shown in Figure 7 of the published article relies on the results of the DNC parsing step 2. Thus, the synthetic chronology is used. The authors do not present any results of canker development in actual time as discovered.

Are the quadrat sampling and related analyses valid?

The published article did not suggest that the quadrat sampling or related analysis were used to support distance of spread estimates. Further, the regression analysis shows only very weak relationships. Thus, the question of validity of the analysis would have little impact on the overall conclusions of the article. However, as documented in Appendix E, and summarized in this section, the statistical analyses based on quadrat sampling is severely flawed and unreliable.

The application of random sampling appears to be highly unusual, because the study sites do not need to be sampled. All properties within the sites were repeatedly surveyed on a regular basis.

The method to calculate representative statistics from the quadrat samples in several cases seemed very odd. One example is statistics related to tree heights. These tree height values were normalized within each sample. For example, if one sample contained heights in the range of 5 to 25 ft, then after normalization, the 5 ft value would equal 0 and the 25 ft value would equal 1. Homeowners' citrus tree heights can vary from about 2 ft to more than 25 ft, so it would be impossible to combine quadrat samples for any meaningful statistics. Additional details are provided in Appendix E. Indexes with normalized sample data are used in 5 of the 10 regression plots in the 2002 article.[8] Thus, there is an unusual allegation in Appendix E that the authors of the published article, purposely "dirtied" their data.

Quadrat sampling extends over "non-citrus" areas and beyond the borders of the site. The contour map shows concentrations of citrus canker in canals, lakes, parking lots, etc. The figure below shows citrus canker located in a lake between NW 192 St and Honey Hill Dr in Carol City.

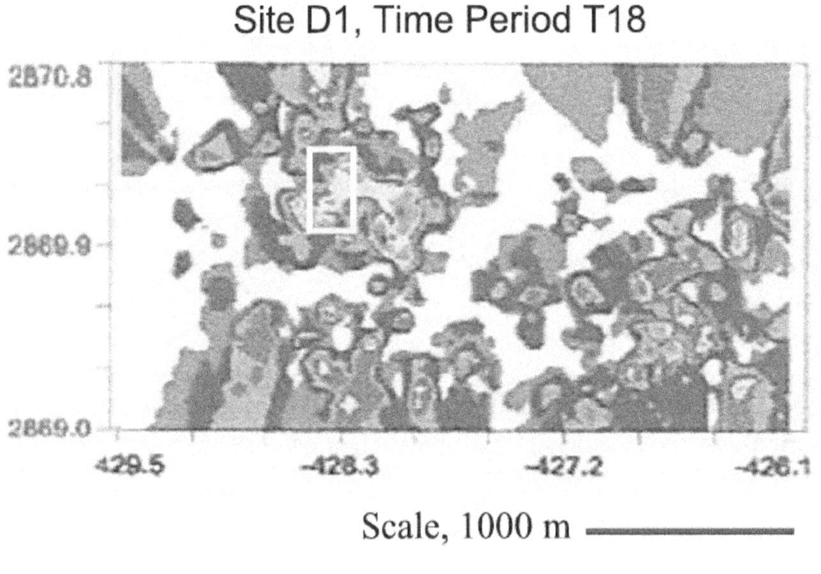

Figure 7.2: Site D1, Time Period T18, with location of lake between NW 192 Street and Honey Hill Dr. , Carol City, Florida

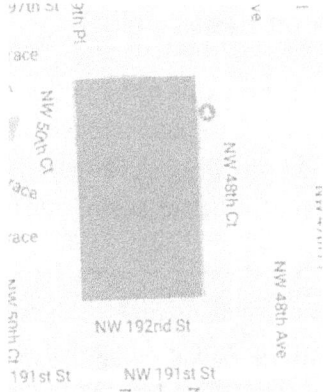

Figure 7.3 Lake at NW 192 St and Honey Hill Dr., Carol City, FL

The image of citrus canker spreading out from a series of foci is simply a creative illusion, as quadrats indiscriminately sample areas which can contain both residential yards and non-citrus areas such as lakes, canals, roads, schools, commercial centers and parking lots. The contouring method, along with the analyses in the next two sections, are based on the assumption of a continuum, where it is assumed citrus canker can be present in any location. While this may be nearly true for a nursery, and a grove with close spacing, it is not true for residential areas. In fact, all the study sites contain large non-citrus areas and there is no real evidence that the bacterium could transverse these areas.

9. Spatial Point Pattern Analyses (Appendices F/F1)

What was done

Spatial point pattern (SPP) analysis provides statistics on a set of points, distributed within a sample space. In the field study, the data consists of the locations of infected trees. It is claimed in the 2002 article that the described SPP method is a modification to Ripley K evaluation and would eliminate the need to consider boundary effects. However, this claim is not supported by any discussion within the article. The statistics are actually calculated by inter-point distance (IPD) analysis. In this case, IPD analysis was performed for each time period. These time periods are a result of the synthetic chronology as discussed in Appendix B.

For each time period, two empirical distributions were developed, an infected tree IPD distribution and a comparison distribution, denoted as Kexp, These distributions are shown in Figure 4 of the published 2002 article.[8] The Kexp distribution is bounded by lower and upper set of confidence limits. While the IPD distribution of infected trees is clearly defined, the Kexp distribution is not.

Are the spatial point pattern analyses valid?

The spatial point pattern analyses do not provide meaningful inferences of the inter-tree distances of transmission of citrus canker bacteria. As explained below, it is also concluded all citrus trees in the backyards of residents are likely to show departures from a complete spatial randomness pattern (CSR) undistinguishable from CSR departures of infected trees alone.

All distance related statistics are based on a comparison of two cumulative distribution as described in Figure 4 of the 2002 published article.[8] The article did not provide clear definition of the Kexp curve, but the text of the article suggested it was based on a complete spatial randomness (CSR) pattern. This conjecture was supported by Monte-Carlo simulation. As described in the Appendices F and F1, a Monte-Carlo model was created to simulate inter-point distances of a CSR pattern, based on a rectangular area. By varying aspect ratios in a trial and error manner, a close match of the Kexp distribution was obtained. Thus, it appears the Kexp curve was based on a computer generated result, rather than field collected data.

The match only added confidence of the basis of the Kexp curve, not the confidence interval. The confidence intervals presented in Figure 4 are different from those obtained from the Monte-Carlo simulation for Site D1. The derivation of confidence intervals based on the hyper-geometric distribution, appears to be in error.

The authors consider the distances related to maximum departure of infected trees from a CSR pattern to be significant. It is not, because all citrus trees, healthy and infected, are likely to show some departures from a CSR pattern. This is based on the simple observation that homeowners are more likely to plant citrus trees in the backyards of their lots. There are no citrus within the numerous areas, such as lakes, canals, roads and houses. Residents do not chose locations at random. Finally, in order to make a valid analysis, the authors would first have to present evidence that the pattern of residential citrus trees can be represented by a CSR pattern.

Since these disparity between curves can easily be attributed to the typical pattern of residential fruit trees found in residential areas, no meaningful inferences can be made to the inter-tree transmission distances of citrus canker bacteria.

10. Semivariance Analysis (Appendix G)

What was done

Semivariance methods are used to identify specific patterns or trends in spatial data. The methods provide statistics on the spatial variation of a specific attribute denoted as a *T* variable, which is a continuous and real variable. Semivariance analysis is used extensively in mining applications, where the attribute is commonly is the mineral content of a unit sample of rock.

In the case of the field study, the *T* variable is the duration that the infected tree was infected according to the synthetic chronology. The synthetic chronology is based on back dating trees to the exact date of their first infection with values t range from 1 to 25 corresponding to their time window. If a tree has an initial infection date (IID) less than the start date of the first period, then $t = 25$. Similarly, if the citrus tree is deemed initially infected in the second period, then $t = 24$.

The semivariance analysis is dependent on the parsing of infected trees in the DNC method. resulting in a synthetic chronology. The transitional spherical model was fitted to the data. From this fit, the properties of the variogram (range, nugget and sill) were obtained.

Is the Semivariance Analysis valid?

The RSTD values were obtained from the theoretical variogram models, not the field data. However, no graphs are presented showing the fit of data to the models, so range values can not be evaluated. Due to the large non-citrus areas, the assumption of a continuum for the transitional model is violated. So, the attribute. *T*, or any other property from the infected trees can not be evaluated using the semivariance analysis.

However, if one considers the RSTD results are valid, it is noted that these RSTD values do not support a conclusion that the citrus canker can be disseminated long distances.. The five RSTD values given in the article range from 9 to 119 m, which is closer to the existing 38.1 m (125-ft) eradication policy, than 579 m (1900-ft) , the post Jan 2000 policy.

However, Dr. Dixon, Bureau Chief of the Plant Pathology section of FDACS/DPI made a presentation in 2001 in public hearings in support of an amendment to Department rules, in which he stated these results collaborate the distances of spread.[2]

11. Field Study Provided No Meaningful Results

Six analyses from the field study were reviewed in detailed as presented in the appendices and briefly summarized in the prior sections. These were not six independent analyses. Each analyses depended on the DNC procedure which created the synthetic chronology based on the discovery dates and lesion ages. In fact, it is impossible to determine from the 2002 article [8] when the discoveries of citrus canker were made, as all analyses are conducted using the synthetic time chronology.

After the determination that the DNC method was invalid, the review could have ended at this point. Instead, each analysis was reviewed on its own, and in each case, the analysis was invalid. So, even if one believes that the DNC method could produce a correct synthetic chronology, the rest of the statistics lack credibility.

Thus, it is concluded that the Florida field study provided no useful information for the determination of the 1900-ft rule. This is actually consistent with the principal investigator, Dr. Gottwald's claim in Broward Court in November 2000 that the general agreement of the 1900-ft radius was not based on any report. It is contrary to the claim of FDACS that the 1900-ft rule originated with the field study results.

The investigation showed each of the six analyses was invalid. This includes the 107 day maximum visibility (Appendix C), the inter-point distance analysis with a maximum distance of 3 miles (Appendix D), and the quadrat analysis demonstrating the spread of citrus canker over lakes, canals, parks and other non-citrus area (Appendix E). This was a surprising result, as from the onset of the review, it was thought there were would be some valid statistical results.

The study lacked basic controls. Obviously, a residential area is not a grove. There are no controls what each owner can do to frustrate the discovery of the disease. The most obvious is removal of infected foliage and fruit. But, citrus canker promotes early leaf drop. A poorly maintained citrus tree would likely have multiple problems — improper planting, lack of irrigation, nutrient deficiencies, and myriad pest and diseases. So, how is it possible, that the single most essential information, the age of the oldest lesion on the tree, could be determined? And yet to avoid errors of incorrect pairing of trees, which directly affect the calculation of distance, this lesion age would have to be determined in terms of day from the initial infection date.

This study is characterized as a hoax or fraud, if one considers this research to be either forming the basis of the 1900-ft rule or supporting its validity. There is no other word which really fits. Viewed strictly from the outside, the 2002 published article[8] on the field study has all the normal trappings of a scientific article. Tables of results, graphs, maps, equations, and even a derivation, which did not make much sense. However, when viewed in detailed, every analyses had major problems. The 107 day maximum visibility as investigated in Appendix C is the result of a statistical trick to produce the appearance of excellent correlation, when in fact, the correlation was poor or non-existent.

It is not suggested that all six authors of the study, as identified in the 2002 published article[8] , understood that the statistical analyses were invalid. The only researcher with expertise in epidemiology, besides Dr. Gottwald, was Dr. Frank Ferrandino. It is believed that Dr. Ferrandino's participation in the study was very limited. He is credited to developing a computer program to provide statistical analysis of the spatial point pattern, but it is likely from the published article, this program was never actually used.

It is believed that the other four authors (Drs. Sun, Graham, Taylor and Riley) of the 2002 article had only general knowledge of the study. While the expertise in the pathology of citrus canker of all four authors is beyond question, none of these scientists appear to have extensive background in spatial analysis and epidemiology.

Nothing within the 2002 article suggested the 1900-ft rule was a direct result of the field study. The article was not determinative of the rule, but supportive of a rule in excess of 125-ft. It is stated that many of the distances of spread as calculated in the study exceeded 1900-ft. The article would be useful in a superficial sense, to demonstrate to the public, the media and judges involved legal challenges to the program, that considerable study had been conducted, and this study was supportive of the current policy.

The question arises, that if this study is so extremely flawed as to offer no meaningful result to regulators, then how could it have been published in the prestigious journal of Phytopathology, published by the American Phytopathological Society. As noted in Chapter 6, the general definition of epidemiology is the change in disease in groups over time and space, not the study of how to control or eliminate diseases. A novel method of analyzing survey data may be considered valuable and publishable research. Also, it is likely the reviewers likely did not give close scrutiny to each analysis, because the article drew only very general conclusions from these analyses. However, the reviewers should have insisted more detailed information on data collection, the location of sites, the data collection period and how certain data such as oldest lesion age, were determined.

12. Back to Property Inspections — Another Piece of the Puzzle

This review is not yet complete. It was concluded that property inspections were at times, incomplete as not all citrus trees or citrus canker were discovered. Subsequent visits may have documented information which was inconsistent with prior inspections. It was presumed that the same FDACS employees conducting routine CCEP inspections would bring GPS meters with them to the study sites, when they did the special surveys within the study site. They would fill out both the normal inspection forms and special forms to be used in the field study. The special forms would be turned into the USDA as Dr. Gottwald would tabulate the results.

The locations of infected trees in study site D2 in North Miami are shown below in Figure 7.4 as provided in the article by Neri et al.[10]. Dr. Gottwald is a co-author of this article and provided the field data. This location is documented in Appendix A as provided on the website. According to the 2002 published article[8], the boundaries conform to a square mile section, as defined by the TRS system (township-range-section). The eastern boundary is North State Rd 9, which is less than 100-ft from I-95, and acts essentially as a feeder road. Yet, infected trees are shown approximately 500 to 700 ft to the east of US I-95. This distances can not be explained by errors in GPS meters.

This is a surprising result, since it is impossible for surveyors to be unaware of which side of US I-95 highway they are collecting data. This is a high speed, interstate highway, with 5 lanes of traffic in each direction (Figure 7.5). The highway would likely act as natural barrier to the transport of citrus by windblown rain. Why would any researcher extend the inspections to the eastern side of US I-95? On the eastern side, these trees are all within residential sections. The nearest infected tree could easily be just over the neighbor's fence and outside the site. In this case, the infected tree would not be part of the study.

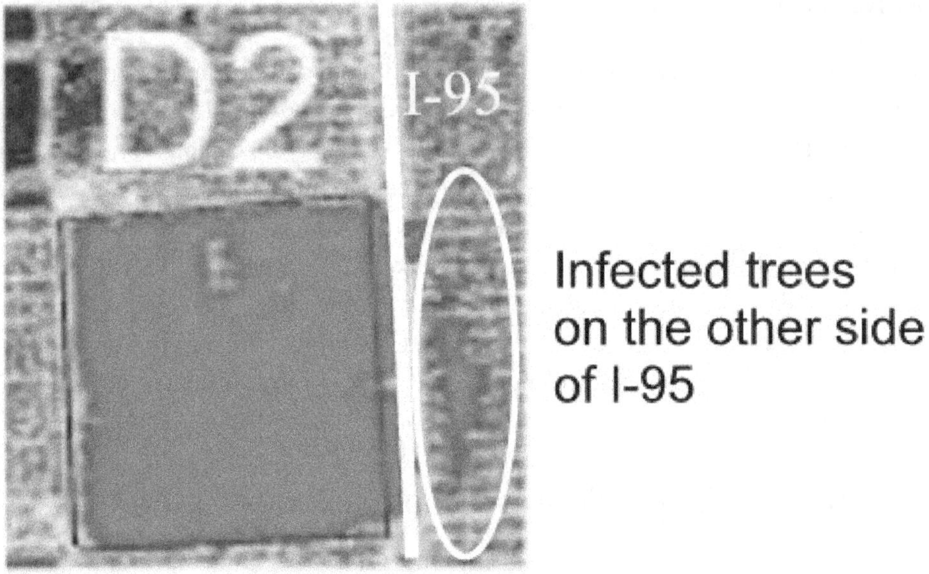

Figure 7.4: Incidences of citrus canker outside of site boundaries (Site D2)[10]

The article in which Figure 7.4 appeared, was published in 2014.[10] It was verified the data are the same because the disease incidence progress curve, as presented in the 2002 article[8], matches exactly with the 2014 article's curve. A similar analysis was done for Site D1 which had an identical progress curve as well.

Figure 7.5: Northbound traffic on Interstate Highway I-95 at Rush Hour in Miami-Dade County

(Source: By B137 - Own work, CC BY-SA 4.0, https://commons.wikimedia.org/w/index.php?curid=48998674)

Also, in study site D1, infected trees were located outside of the site's boundaries in every direction. As described in Appendix A, the infected trees in site D1 are located from 200 to 700 ft beyond the major roads which presumably were the site borders.

More Discrepancies and Uncertainties Identified

Discrepancies in boundary locations from various presentations are noted in five study sites as described in Appendix A. For example, the size of study site B1 is stated as 6.0 square miles based on two sources, while three others sources show it to be 2.75 square mile, approximately half the size.

As part of this investigation, each site was visited. Each bounding street was a major road. It was unreasonable to assume surveyors would not know which were the confining streets and stay on the proper side. Also, the parcel lot number for each property was easily available from the CCEP database, providing the Section-Township-Range identifier. Surveyors making repeat visits would know when they were inside or outside of the site.

There is uncertainty of when data were collected. The published article[8] provides no dates for the end of data collection. Correspondence from the Department varies anywhere from June to November 1999 as provided in Appendix A. Based on the published article[8], the end date of the last temporal window was November 14, 1999. But Dr. Gottwald sent to the Department an interim report a month earlier, on October 13, 1999 — a month before the end. What sense does this make? Further, what is the sense of obtaining general agreement on the 1900-ft rule in a meeting in the December 1998, if the field study was slated to last one year (through at least March 1999, 12 months since Commissioner Crawford announced the field study).

13. Department/USDA field study narrative starts to fall apart

The conflicting information comes from the very best sources, publications by Dr. Gottwald of the USDA/ARS and a letter from the Director of FDACS/DPI, Mr. Gaskalla. The inspections of backyards during the field study were at the direction of the Department. How could they not know where the study had taken place?

It seemed unrealistic to believe FDACS sent Dr. Sun to "visually inspect" all citrus trees infected with citrus canker. He would have to practically live in South Florida. If there were 3323 infected trees within the sites, there were likely many be more in the surrounding area. If a single one mile square section is inspected in the surrounding area, just ½ mile from the boundary, approximately three square miles would have to be inspected.

The 2002 published article on the field study and the 1999 interim report discuss the conversions of the longitude and latitude data to UTM coordinates. The published article identifies two models of GPS meters. But, the GPS Garmin model 12XL had an option to read out in UTM coordinates. Why bother with conversions? This just did not make sense.

Then at the end of the study, not a single document on the study was retained by FDACS. The entire dataset would have likely fit on a single CD, even in year 1999. This was very odd considering it was their inspectors checking regularly on the health of approximately 19,000 citrus trees. Why would the Department not be interested in retaining the information gathered in the study on the age of the citrus tree and the size of the canopy, for all 19,000 citrus trees?

It has already been mentioned how odd it was, that inspectors did not check for citrus leafminer. The Department emphasized at every opportunity, how injuries by the leafminer would make citrus trees many times more vulnerable to canker. Symptoms of citrus leafminer would be very obvious to inspectors.

Assuming the decision was made in December 1998 to use 1900-ft, then why would the Department want to continue to send inspectors into the same study sites from January to November 1999 to repeatedly inspect the same properties every 60 days? Why would the Department want Dr. Sun to inspect visually the infected trees from January to November 1999? Funding was scarce, and one would think inspectors during 1999 would be needed in the northern areas of Broward and Palm Beach Counties, instead of the three sites within Miami-Dade County and two sites in Broward, close to the Miami-Dade/Broward County borderline.

It is further not explained how inspectors on their 60 day return visit to the properties would be able to correctly locate the same citrus on the properties. Due to the inaccuracy of GPS meters, they would have to rely on some notes to be sure they had the right trees in cases of more than one tree. Would they again record the lesion ages, the variation of canker lesions, top to bottom, or directionally, north, south, east and west? Would they again record all information? What would they do if the data were inconsistent with the previous visit?

Further, read carefully, the 2002 article does not say that GPS meters were used every day in each site. It does gives some specifications of two models of GPS meters, with the implications these were used during the study. Also, FDACS webpage discussion of the "Epidemiology Study" states that only GPS technology was used.

A Desktop-based Field Study?

It is believed the entire field study was done simply for political and legal reasons. Prior to the creation of the Task Force in February 1999, the 1900-ft policy decision had already been made between the two key participants— USDA/APHIS and FDACS/DPI. This is consistent with Dr. Gottwald's court testimony that there was already a consensus by December 1998, to extend cutting to 1900-ft. He also states this in the 2001 article.

A possible explanation for the conflicting information on the study sites, and the lack of information on the start and end dates is the field study was not conducted by a special group of inspectors with GPS meters. There may have been a few special inspections in the beginning, but not for the entire study period.

It is alleged that most or all of the field study was done in Dr. Gottwald's office at the USDA. More specifically, Dr. Gottwald simply downloaded prior inspection records. Much of the information would be of no use to him, such as the names and addresses of the residents. But, the CCEP database included longitude and latitude of every property. The longitude and latitude data was used by CCEP to calculate eradication radii.

Possible use of CCEP database data to locate citrus trees

It is suggested the center locations of residential lots came from the CCEP database instead of actual tree locations in the Florida field study. The center locations were part of the CCEP database since it was necessary data to determine the removal distances.

The USDA/ARS refused to disclose all data involved in the Florida field study, including the locations of citrus trees. If the CCEP database were used, this could be seen in the data, as every citrus tree within a particular lot would have exactly the same coordinates corresponding the lot's centroid location.

But would accuracy really have been lost if the CCEP database was used? It is likely at times the database locations could be less prone to error particularly with smaller backyards. In examining lot sizes, many lots have backyard spaces approximately 1,500 to 3,000 ft^2. Assuming a square geometry with an area of 3,000 ft^2, this space would be 55 x 55 ft. Also, citrus trees are likely planted from 10 to 15 ft from the back of the yard, the trees would be 30 to 45 ft away from the house. In the published article in 2002, and others presentation prior to this article, the GPS meters are accurate to approximately +/- 25 ft. Given the small area likely to contain a citrus trees, it is fairly easy for GPS meter-based coordinates to place the citrus tree in the wrong lot. This is particularly true if the citrus tree is located in the corner of a lot. At least, if the CCEP database was used, these errors would not occur.

Oldest lesion age — all fun and games

It is alleged that the "oldest lesion age" identification, through backyard visits were either totally or partially fabricated. Perhaps, Dr. Sun made a few trips to Miami, so there would be some record in case he was required to testify. Many of the lesion age determinations were not done by any of the other inspectors, but were created by Dr. Gottwald as he sat at his computer, putting data into Excel spreadsheets. It would not have taken him a long time. As the lesion ages were created by Dr. Gottwald, the locations of PI and NI trees in each hypothetical scenario would be displayed. At this point, it was all fun and games.

However, it is suggested that data fabrication had some limitations. The Excel spreadsheet contained only infected trees were identified by Department inspectors. No infected trees or addresses were ever invented. It is also possible that Dr. Sun visually identify some of the infected trees in Miami at the beginning of the study.

The "Desktop" field study would explain plenty

If this desktop field study narrative is true, it would explain many of the more puzzling aspects of the study. First, is the secrecy and confusion surrounding the inspection dates and site locations. The missing details on the study were very obvious. It would explain why FDACS would claim it retained none of the field data. Imagine a state agency which spends 18 months collecting field data, and has no desire to retain any data. It would explain the extraordinary efforts of the Department to avoid a trial on the validity of the 1900-ft rule, which would require depositions from everyone, all the way down to those conducting the surveys.

The desktop field study would also explain how, if data collection ended by July 17, 1999, as Dr. Dixon stated in the Public Hearing, it was possible for Dr. Gottwald to present data extending to November 15, 1999. The answer was he did not need to collect the data, he just downloaded records from the CCEP database.

It is possible explanation of why the only date, ever mentioned in published articles is August 1998, the date of a cooperative agreement, because this is the date that Dr. Gottwald is given full access to the CCEP database. It is possible that on the date of this agreement, it was understood by Dr. Graham of UF/IFAS and Drs. Dixon, Sun and Schubert that the field study would be a desktop exercise.

The desktop field study would explain the selection of Carol City, in Miami-Dade County as Site D1 where access to property was the most difficult due to chain link fences. It always seemed very difficult for Dr. Sun, fairly short in physical stature, could access back yards surrounded by chain link fences.

The age of each tree and the size of the canopy were part of the collected data, but never used in any of the analysis. The authors never state how the tree age was determined. Data were either created on the desktop, or absent altogether.

The presence of citrus leafminer presence was not collected because this information was not within the CCEP database. Did officials ever wonder why the field study did not include the presence of citrus leafminer?

It would also explain the strange procedure used in the creation of indexes for correlation analysis. Researchers frequently take pains to "clean up" data to make it more meaningful. However, in this case, the strange normalizing procedure seemed to actually "dirty" the data, making indexes and the related correlation analyses less meaningful. The discussion is provided in Appendix E. It is conjectured that Dr. Gottwald did not want any meaningful results to come from improperly collected data on approximately 19,000 trees.

Does it matter?

While the allegation that some of the Florida field study data were fabricated may seem very serious, perhaps this is not as serious as it seems. It has already been established that none of the six analyses are valid. Might they be valid, if data collection occurred on the ground, rather than data downloaded from the CCEP database? The answer is an emphatic no — there is not a single statistics within the field study which would be improved if the data unique to the field study were collected by surveyors.

It makes little difference if an inspector goes to a tree, and guesses at the oldest lesion age, or Dr. Gottwald makes up the numbers in his office. In fact, it would have been a waste of time, making repeated visits to back yards, if the end results were going to be the same.

14. Concluding Remarks

The conclusion reached in this chapter is every single analyses in the Florida field study as published were so severely flawed that neither the statistics nor the conclusions are meaningful to the establishment of an eradication radius. To the layman, the articles have the look of science, full of equations, tables, and figures. Under close examination, there was not a single analysis which stood up against scrutiny. Perhaps, the most telling of all statistical analyses was the weather analyses. The use of cumulative sum variables was simply an old statistical trick. This led to the conclusion that the entire field study was done for show.

However, this conclusion then turns to a question of how could a group of scientists from the USDA, the FDACS and University of Florida/IFAS collaborate on such a hoax. How is it possible that Dr. Gareth Hughes, a well known epidemiologist, and co-author of a recent textbook on plant disease epidemiology[9] would be part of a study that was seriously flawed? It is suspected that this "team effort" never existed and the 2002 article was written entirely by Dr. Gottwald. From the end of the study in late 1999 to the submittal of a manuscript in November 2001, there was nearly two years to prepare the article.

The Department has contended that the 2001 and 2002 articles were validated through the peer review process. This is the normal process of selecting submitted manuscripts to be published. It is not a thorough review of methodology and results. Further, it was stated the 1900-ft policy was recommended at the May 11, 1999 meeting of the Risk Assessment Group. The only reporter at the meeting, Mr. Paul Power from the Lakeland Ledger did not report any change in policy in his May 12, 1999 article.[11]

It is alleged in this chapter that much of the data were not collected, but simply downloaded from the CCEP database. Further, it is alleged that some of the data were simply fabricated, but the significance of this minimal, because the entire study was meaningless to the policy decision. It however, resulted in the USDA/ARS refusal to release their data, even after a Freedom of Information Act request had been submitted.

It is believed Dr. Gottwald was being honest, when he stated that the 1900-ft rule was not based on any report. However, Dr. Gottwald's assertion that the 1900-ft distance was the result of an unofficial meeting in December 1998 at USDA/ARS office with officials of various state and federal agencies and citrus industry leaders and where no minutes were recorded nor was there a list of invited guests seems highly implausible.

Another piece will be set in place in Chapter 8. Please keep reading.

Short Notes on the Website:

SN 1.3 Long Distance Transport of Citrus Canker by Hurricanes and Tornadoes
SN 6.2 The 1900-ft Policy and the Published Articles by Dr. Gottwald et al.
SN 6.3 Comparison of Department Website Justification Statements
SN 6.4 1990 Highlands Observational Study

Key References:

7. Gottwald, T.R., Hughes, G., Graham, J.H, Sun, X., Riley, T., 2001. The Scientific Basis of Regulatory Eradication Policy for an Invasive Species, Phytopathology, 91:30-34.

8. Gottwald, T.R., X. Sun, Riley, T. Graham, J.H., Ferrandino, F. and Taylor, E., 2002. Geo-Referenced Spatiotemporal Analysis of the Urban Citrus Canker Epidemic in Florida, Phytopathology, Vol 92, No. 4.

Appendices:

These appendices provide the critical review of the Florida field study. Readers' comments may be included in Appendix H, or posted separately on the website. Comments will only be posted with permission of the sender.

Appendix A: Basic Information of the Florida Field Model
Appendix B: Distance Necessary to Circumscribe (DNC) Method
Appendix B1: Unusual Field Study Results
Appendix C: Weather analyses
Appendix C1: The Gottwald Canker Forecast Model
Appendix D: Inter-Point Distance Analyses
Appendix D1: IPD Supplemental Information
Appendix E: Random Quadrat Procedure and Related Analyses
Appendix F: Spatial Point Pattern Analyses
Appendix F1: Supplemental Information
Appendix G: Semi-Variance Analyses
Appendix H: Additional Epidemiology Review/Comments/Errata*

* Appendix H will be added after publication and updated as needed.

8. UNDISCLOSED STUDIES

> The implementation of the 1,900-ft rule will result in an effective clear-cutting of the majority of dooryard citrus within approximately 793 square miles (2,054 km^2) of the Miami metropolitan area in Dade and Broward Counties. An estimated 750 thousand dooryard trees will be removed from urban areas in Dade and Broward Counties within the next year.

Gottwald, T. R., Hughes, G., Graham, J. H, Sun, X., Riley, T., 2001, The Scientific Basis of Regulatory Eradication Policy for an Invasive Species, Phytopathology, 91:30-34

---

1. Filling In the Gaps

Chapter 7 made a critical discovery — the Florida field study was not the origins of the 1900-ft rule. When carefully reviewed, the study as presented in the 2002 published article[11] did not provide any meaningful analyses to establish an eradication radius. Further, the article was published more than two years after adoption of the 1900-ft policy. Thus, one piece of the puzzle is solved. Yet, if 1900-ft did not come from the field study, then what was the true origin of the rule?

As previously mentioned in Chapter 1, Section 8, Dr. Gottwald in sworn testimony on November 9, 2000, in Broward District Court when asked where the 1900-ft distance came from, stated:

> No, the report does not discuss the meeting that occurred in December 1998 in which there was a group of scientists, regulatory agents and growers present in a room and decided upon 1900 feet. 1900 foot *[rule]* is not decided upon this, the manuscript or anything else, it was decided upon by a group, a body of regulators, university scientists as well as ARS scientists, etc. It was not decided upon by these reports.

Dr. Gottwald was referring to an interim report, sent to the Department on October 13, 1999.

Dr. Gottwald is being partially truthful, the 1900-ft rule did not come from any report. However, given the enormous impact of the change to a 1900-ft rule, the choice was not done in an arbitrary manner, by an unnamed group assembled at the USDA/ARS office in December 1998. At the Broward Court trial in November 2000, the Department could not provide a single document to show who attended the meeting and what was discussed. In fact, the December 1998 meeting may have never taken place or it was simply a meeting of the inner circle of plant pathologists and two officials, Deputy Commission Craig Meyer and Director of DPI Richard Gaskalla.

Evidence suggests that a computer simulation study was developed by Dr. Gottwald and likely completed prior to December 1998. It is believed that this was the true origins of the 1900-ft rule.

The model provided assessments on the areal coverage and tree eradication estimates for various radii. A healthy to infected tree (H/I) ratio for each radii could be calculated.

Incredibly, the field study was entirely a smoke screen. Remember, in all communications and publications, neither Dr. Gottwald nor the Department ever used the words, "field study", but always opted for a more vague wording, "epidemiology research", or the "Gottwald study", which could include other research studies. The epidemiology research identified "distances of spread", which again was vague. Epidemiology is a broad discipline, and includes both conceptual simulation models and weather analysis as discussed in Chapter 6.

It was unlikely that a team effort was involved in the field study. All presentations were made by Dr. Gottwald, usually accompanied by Dr. Graham, UF/IFAS professor and soil microbiologist. Dr. Gottwald was the principal investigator for the citrus canker epidemiology research, and no other scientist in the USDA/ARS or USDA/APHIS was part of any study. He was the first author on the two publications related to the field study. Others were included at participants, but their participation was more as co-authors to presentations made by Dr. Gottwald. The Department stated that all data related to the epidemiology research was retained by Dr. Gottwald.

Dr. Gottwald did not chose the 1900-ft radii. Rather, he provided the technical information by which officials could use in selecting a radius in less than a totally arbitrary manner. It was founded on one truism — if 100% of all citrus were eliminated from an area, then there would be 100% elimination of citrus canker. It would follow that any radius that destroyed a high percentage of citrus trees, would effectively eliminate the chance for subsequent outbreaks of canker.

In addition to simulation studies, it is suggested that there was a second series of undisclosed internal studies were conducted from 1996 to 1998. These studies involved a field experiment in Miami-Dade County to demonstrate the potential distance of citrus canker transmission. The supporting evidence of the field experiment is speculative, and only briefly reviewed as it had no impact on the eradication program. The researchers in this field experiment were Drs. Gottwald and Graham.

2. Evidence leading to a Simulation Model Approach

The following are three key facts which lead to the conclusion that the assessment of the 1900-ft and other radii was the result of a simulation model approach developed by Dr. Gottwald.

#1: The number of trees to be eradicated needed to be estimated

Any recommended eradication radius would need some estimate of cost, which in turn would require an estimate of the number of trees to be eradicated. As the larger radius are evaluated, the chance the eradication circles will overlap must be considered. This is not a trivial problem. The number and locations of future infected trees would be impossible to identify on a map. The most effective way to estimate the "clear-cut" area (areal coverage) including an overlap factor would be using a computer simulation model based on random number generation. The model is explained in detail in the next section. The same computer model used to calculate the percentage of areal coverage for any specific radii could estimate the number of trees eradicated for a specific radius. Thus, both areal coverage and tree eradication estimates (healthy and infected) would be determined by the same computer program. The program would also determine the healthy to infected tree ratio.

Dr. Gottwald presented the eradication estimate of 750,000 to 1,500,000 trees in Miami-Dade County in November 2000 (Figure 8.1) It is likely this is the same presentation presented in June 2000 at the International Citrus Canker Research Workshop.

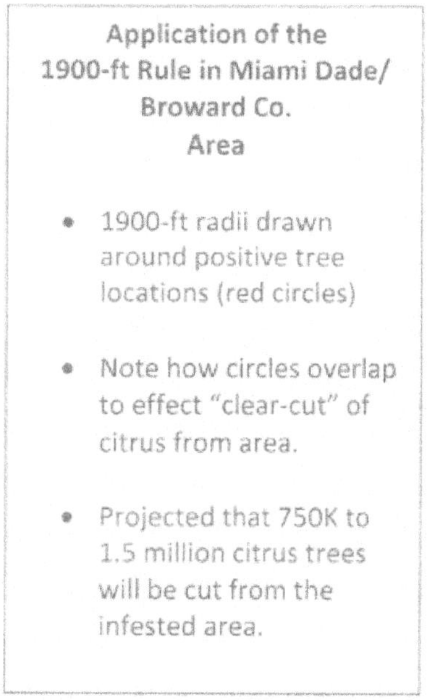
Application of the
1900-ft Rule in Miami Dade/
Broward Co.
Area
• 1900-ft radii drawn around positive tree locations (red circles)
• Note how circles overlap to effect "clear-cut" of citrus from area.
• Projected that 750K to 1.5 million citrus trees will be cut from the infested area.

Figure 8.1: Presentation by Dr. Gottwald, Broward Court, November 2000

It would make little sense for officials, including as Mr. Craig Meyer, FDACS Deputy Commissioner of Agriculture and Dr. Stephen Poe, Program Director for Citrus Canker, USDA/APHIS to discuss various eradication radii without budget and general logistics estimates, in particular the number of cutting crews and program duration. This would require the tree eradication estimates.

#2: The necessary simulation program had already been developed

Dr. Gottwald created essentially the same computer routine that would provide the eradication estimates. This was revealed in the April 2002 article which provided details on the field study.[11] The simulation model as used for areal coverage estimates would randomly locate circles of a specified radius within an area of interest (square mile section). This is essentially the same as the random quadrat sampling used in the Florida field study, as discussed in the published article[11] and reviewed in Appendix D.

In the April 2002 article[11], randomly located squares were employed in quadrat sampling while the clear-cut model, used circles. It is likely that Dr. Gottwald first created the program with randomly located circles to calculate the areal coverage and other statistics using Excel Visual Basic Applications (VBA). Then he changed the geometry of the sampling objects from circles to squares. It would require only a minor changes to the code.

As discussed in Appendix D, the quadrat sampling as presented the April 2002 article[11] appeared to be a very unusual application, since all lots within each study site had been repeatedly surveyed. However, it makes perfect sense that the original intent of the program was to calculate areal coverage and the number of trees (healthy and infected) to be eradicated for various radii.

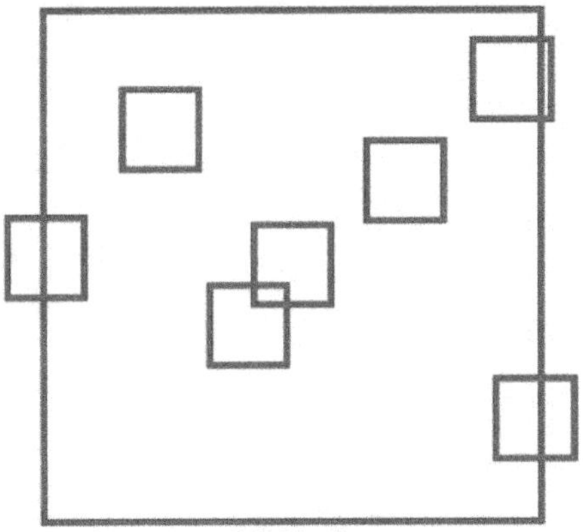

Figure 8.2: Random Quadrat (0.5 km on a side) in a square mile

#3: Determination of a clear cutting radii was the intention of the Voronoi Tessellations - Broward 2000 Presentation

It is suggested that the presentation of Voronoi tessellations indicate Dr. Gottwald was looking for a method to estimate a radii with a high level of areal coverage, consistent with the objective an eradication program which would leave few pockets of citrus in residential areas.

The Voronoi tessellations presentation was presented in Broward court in year 2000, and likely this is the same presentation in June 2000 at the International Citrus Canker Research Workshop.[8] Tessellations refers to the tiling of a flat surface. With this method, the boundaries of the polygon are drawn to be equidistant from the points on either side of the boundaries. An more complete explanation of this technique is provided in the online supporting documents website. Also, many examples can be found on the internet.[14]

The presentation of Voronoi tessellations at the Workshop in June 2000 was very brief. Dr. Gottwald does not indicate why the analysis was done, nor offer any numerical results from the work. It is suggested the method was used to identify the expected areal coverage of various radii. This would be consistent with the goal of clear-cutting a high percentage of citrus within residential areas. This work may have been used to complement the simulation work. The distance from the infected tree to the most distant vertex of the polygons (Voronoi cell) could be calculated for all cells. These distances could then be sorted from low to high and percentile levels could be calculated. The percentile levels would give a probability value for each distance. Additional discussion of Voronoi tessellations can be found in Short Note 8.2.

The results would complement the simulation modeling because this work does not assume a fixed number of trees in each section, and the trees are not assumed to be randomly distributed. The method could be used to calculate the high probability distances for all trees within Miami-Dade and Broward Counties, if tree locations were approximated using Department's database.

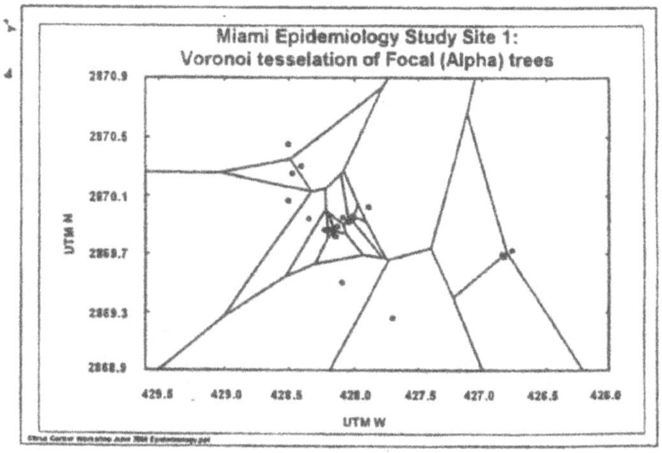

Figure 8.3: Miami Epidemiology Study Site 1, Voronoi Tessellations

In sum, the 1900-ft rule was a mathematical result, solved numerically by simulation. It was not a biological one. The objective was not to capture the next generation of infected trees but to provide a high level of areal coverage with large eradication circles. It would also provide the tree eradication estimate, needed for planning purposes.

3. Monte-Carlo Simulation Model Concepts

A stochastic or Monte-Carlo model uses repeated generation of random numbers (replications) to represent uncertainty in the system. It is a specific numerical solution to a mathematical problem. In the model, infected trees within a specified area can be located with randomly selected coordinates as shown in Figure 8.4. To illustrate the concept of overlap, the circles with a 300-ft radii were drawn around these locations. Each circle would encompass 6.5 acres, so with no overlap, 10 infected trees would cover 65 acres or about 10% of the area would be cut down. But, the areal coverage is less than this, because some circles overlap. Also, areal coverage is reduced because some circles lie partially outside the square,

To calculate the effect of overlap, it is necessary to run the model many times, with different random locations of infected trees and calculate each time the percent of area which would be enclosed by the circles. The multiple runs or realizations provide estimates of expected areal coverage and eradication estimates. Programs such as Excel with Visual Basic Applications can simulate the affect of various radii on the areal coverage.

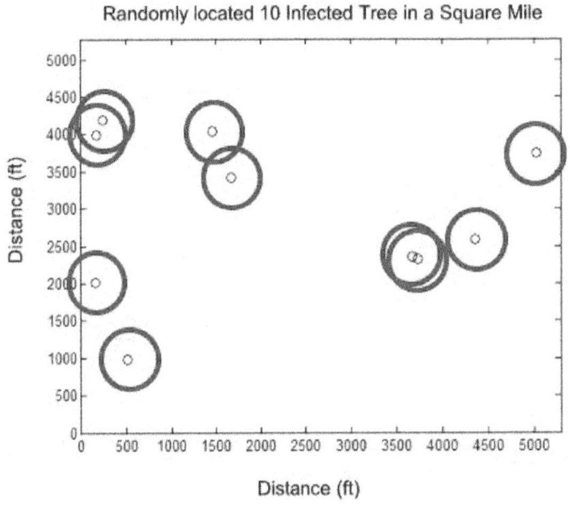

Figure 8.4: 10 Randomly Located Trees in a Square Mile with 300-ft Circles

4. Model Procedures and Results

To test the proposition that the 1900-ft rule was chosen based on a simulation model, two models were constructed using the MATLAB© program. Each model requires two input values: (1) The number of infected trees in the specified area and (2) The cutting radius. The result is the percentage of clear-cut area or areal coverage. This is the equivalent to the probability that any healthy tree randomly located within the section would be cut down. Each model considered 10 infected trees randomly located within a square mile

The first model, identified as the "Closed Model' determined the areal coverage based only on infected trees within the square mile block. It is closed to the effects from other adjacent sections.

The second model, identified as the "Open Model" included the effects of infected trees in the surrounding sections, as shown in Figure 8.5. The center block and surrounding blocks (a total of 9 blocks) are populated with 10 infected trees each, and randomly located. Calculated percentage of clear cut area was calculated only within the center square. As shown below, infected trees located within the eradication radius of the center block will increase the percentage of the area clear cut.

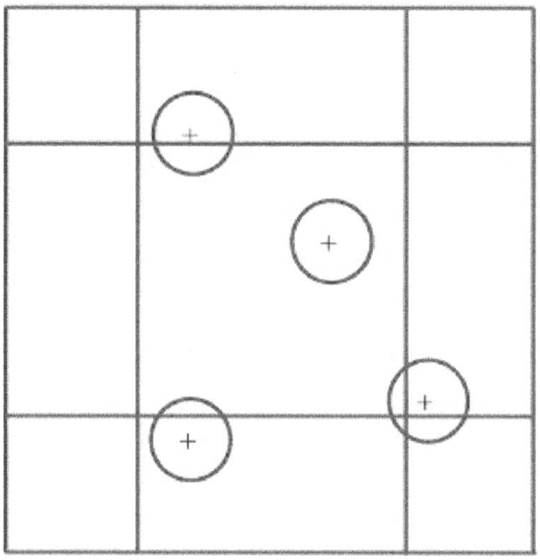

Figure 8.5: Open Model Showing 4 Infected Trees Outside of the Center Section

Areal coverage results

Results are shown in Tables 8.2 and 8.3 for closed and open sections, respectively. All results are the average of at least 3 runs with 300 replications each. The results are accurate to about +/- 0.2%. The accuracy of estimates depends only on the number of replications and the spacing of the mesh points used to calculate areal coverage.

Table 8.2: Closed Section Model Results

Radius (ft)	% area with trees removed
2700	99.5
1900	94.7
1600	89.8
1200	74.6

Table 8.3: Open Section Model Results

Radius (ft)	% area with trees removed
1900	98.7
1600	95.3
1400	89.3
1200	80.8

The "% area with trees removed" can also be considered an estimate of a theoretical probability. It is the probability that any citrus tree within a section, would be located within at least one eradication circle, given 10 trees are infected in the section and both healthy and infected trees are randomly located. It is a theoretical result, as tree locations are unlikely to be correctly represented by a complete spatial randomness pattern.

For 1900-ft, the simulation model results in either 95 or 99% areal coverage, depending whether a section is open or closed. A property which borders on a canal, lake or any large non-citrus area would be partially closed, so the areal coverage would be between these two estimates. If a section has fewer than 10 infected trees, or the infected trees are clustered together, then the areal coverage percentages are lower. Complete MATLAB code and discussion of computer programs are provided in the online supporting documents website (Short Note 8.1).

Eradication estimates results

In both the November 2000 presentation to the Broward Court, Dr. Gottwald accurately predicted the number of citrus trees that would be cut down in Miami-Dade and Broward counties was from 750,000 to 1,500,000 citrus trees. The actual number is 850,643 trees for all counties or 717,016 trees for Monroe, Miami-Dade, Broward and Palm Beach counties. The equation would be:

$$Estimated\ eradicated\ trees\ =\ (areal\ coverage)\ x\ (citrus\ trees/sq\ mile)\ x\ A$$

where A = the square miles in Miami-Dade and Broward counties, residential areas only.

Areal coverage estimates would vary depending on the expected number of infected trees. The average number of citrus trees per section would likely be available from the CCEP database. Only residential sections would be included in the calculations.

5. The Table of Model Results — May 11, 1999 Report

Taking results the computer simulation study and combining it with another one, would be highly unethical, but it is believed this is exactly what Dr. Gottwald did on multiple occasions. Specifically, it is believed that the "epidemiology results" as given in the May 11, 1999 CCRAG meeting report, contained on one line, the results of a clear cutting simulation model, specifically, the values as obtained from Site 1 3rd 1-mo window.[2] The same table of results is presented in the minutes following the May 14, 1999 General Task Force Meeting.

Table 8.4: Comparison of CCRAG-9 Values and Open Section Model (1900-ft = 99%)

Radius (ft)	May 1999 CCRAG Values*	Open Section Model Values
1900	99%	98.7%
1600	95%	95.3%
1200	90%	80.8%

* Per released report from the Citrus Canker Risk Assessment Group based on the meeting, May 11, third period of 30-day results. [2]

The Interim Report, sent to Dr. Wayne Dixon on October 13, 1999, and then forward to Mr. Richard Gaskalla, and Mr. Craig Meyer, shows the same values.[2] These same values were published in the 2001 Letter to the Editor article, by Gottwald et al. in Table 1, 3rd 1-mo. temporal window as submitted to the Broward County Court.[9] Dr. Gottwald's presentation in

November 2000, also state that "CCEP regulatory agency chose to use the 99% level" thus 1900-ft is associated with 99% probability.

Comparisons with other presentations

There were numerous other presentations, where the 1900-ft distance was associated with the 95% probability level and the 2700-ft distance was the 99% level. The statement which appeared on the FDACS website stated:

> From this research *[epidemiology study]*, they determined 95% of the exposed trees that became diseased were up to 95% feet away from a single positive tree... subsequent infections resulting from inoculum dispersal from focal trees lie within approximately 1200 feet 90% of the time, 1900 feet 95% of the time and 2700 feet 99% of the time.

The 1900-ft was associated with the 99% probability level in Commissioner Crawford's letter to President Clinton in year 2000, as provided in the reference section of the online supporting documents website. However, Dr. Gottwald presented the 1900-ft distance as corresponding to the 95% level in the November 2000 Broward Court presentation (Figure 8.6).

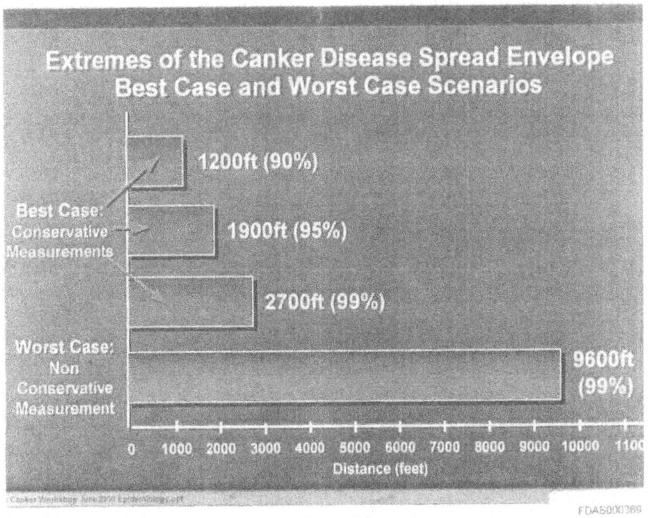

Figure 8.6: Dr. Gottwald Slide Presentation, Broward Court 2000

A comparison of the "website values" with the closed section model, shows close agreement with 2700-ft and 1900-ft, while the 1200-ft values are significantly different as shown in Table 8.5.

Table 8.5 Comparison of FDACS Website values and Closed Section Model

Radius (ft)	FDACS Website Values	Closed Model Values
2700	99%	99.5%
1900	95%	94.7%
1200	90%	74.6%

In simulating the 1200-ft radii, Dr. Gottwald may have included an option to assume that re-inspections would occur in the remaining 25% of healthy trees, and there was a chance that one or more of these trees would be discovered with citrus canker, thus the second eradication effort would result in a higher level of eradication.

It is possible that Dr. Gottwald made these results known through the Risk Assessment meetings. Some members of the group may associated the 1900-ft with a 99% probability level as in the open model, while others associated it with a 95% level, as in the closed model.

6. Chipper Experiments

As part of the 1900-ft policy, the eradicated citrus tree and all plant parts including the citrus fruit had to be processed through a wood chipper. Since the disease is local (non-systemic), the weight fraction of the infected trees with viable bacterial cell is miniscule as compared to the entire tree. The wood chipper produces mostly 1 to 2" wood chips, which would land in the covered truck. Some of the chips and dust may exit the truck, but not travel very far.

The Progress Report on the project, likely submitted in 1999, states that "plant material" placed around the truck became infected. Further, it is stated at the International Citrus Canker Research Workshop, that grapefruit seedlings became infected up to 20 ft from chipper. In an article in 2004, with Dr. Graham as the first author seemed to be backing off from these results and stated that air samplers had identified some "viable cells" near the chipper. Thus, with each publication of the project, the story seemed to change. The full discussion of the chipper experiments and alternative explanation is provided in Short Note 8.3.

It is likely that a containment greenhouse had been parked for some time at the Opa-Locka airport. However, it is believed the rest of the chipper story, as told at the Research Workshop, was a cover story for some field experiments on the potential range of canker dissemination. It is likely that the experiments could not demonstrate inter-tree transport beyond a relatively short distance, so they were never disclosed. If experiments were conducted as is theorized, they had no effect on the program.

7. Concluding Remarks

It is alleged that there were undisclosed studies. The first study was entirely done on the computer. The assessments helped officials decide on the 1900-ft radius. The second study was a field experiment which had no affect on the program. Short Note 8.3 on the website provides more details on "chipper experiment" which is considered not to have been at all focused on chipper dust.

The first study identified areal coverage for various radii using computer simulation model. These results were unrelated to the natural transport of canker by windblown rain. The model is entirely based on mathematics, not biology. In this case, Monte-Carlo simulation provides a numerical means of solving a mathematical problem. However, could the computer studies as suggested here be considered good science? The answer is yes. Had they been presented, it would have fit all the criteria for good science. It was completely reproducible. It could be easily described. Finally, it was useful epidemiology information as the model could accurately estimated the number of trees to be eradicated.

The problem was not in the result, but the question it answered. The model simply stated that any random scattering of 10 points in a square mile, with 1900-ft circles drawn around them would, on the average, cover 95% of an area. The stated objective of the eradication program was not supposed to clear-cut large areas in urban residential areas, but eradicate the next generation of infected trees. After January 2001, there would be no mention of clear cutting in any article published by Drs. Gottwald, Graham and Schubert.

The best evidence of these undisclosed studies comes from the necessity of the program to have estimates of the number of trees to be removed in residential areas. This in turn, requires an estimate of the areal coverage. It is not a trivial problem, as the circles would overlap.

Further clues of an undisclosed model were from Gottwald presentation of quadrat sampling in the April 2002 article and the Voronoi analysis as presented in the Broward Court in year 2000. Since neither of these analyses seemed to contribute to the presentation of epidemiology research, it is an open question of why Dr. Gottwald presented these analyses.

Short Notes on the Website:

SN 8.1: MATLAB routine to calculate areal coverage
SN 8.2: Further discussion of Voronoi Tessellations
SN 8.3: Chipper experiments and alternative interpretation

9. A NEW HISTORY EMERGES

> "We've made a monster out of citrus canker, " Griffiths said. The Argentines "are living with canker ... very successfully."

Jim Griffiths, General Manager of Citrus Growers Association. Orlando Sentinel, May 24, 1987.

> ... the political pressure exists for the eradication effort to go ahead with the ambitious goal of removal of all diseased and exposed citrus from Miami outbreak area within one year.

Based on a manuscript of a Letter to the Editor of Phytopathology by Gottwald et al, as submitted to the Broward Court in November 2000. This comment was later removed.

---

1. Finally, The Puzzle Comes Together

We have arrived. It is time to reassemble the puzzle, to re-state history consistent with what has been demonstrated to be true or what are the most logical explanations of events. Also, it is time to discard what is likely to be untrue. A false statement can not be made into a true one by repeating it a thousand times.

This chapter presents the best explanation for the events leading to the 1900-ft policy and then later to the widespread dissemination of citrus canker throughout Florida. This is really the only comprehensive narrative of events leading up to the 1900-ft rule. The Department story is typically contained in a short single paragraph.

Our journey began in Chapter 1 with the recognition that the research on the 1900-ft rule seemed odd and inconsistent. For example, a 2001 Letter to the Editor in Phytopathology showed a distance of spread of canker up to 11 miles.[8] A year later, a new article by essentially the same authors showed the distances of spread were all far less with a maximum distance of 2.2 miles.[9] Many other inconsistencies were noted in Chapter 1, including large changes in the number of citrus trees within the study sites from one presentation to the next.

The means of dissemination were reviewed in Chapter 3. Many of transport mechanisms, such as canker spread by birds were without scientific evidence.[5,6] Selected topics in epidemiology were reviewed in Chapter 6. Basic terms were defined including observational and experimental studies. A review of the literature showed the maximum distance of dissemination by windblown rain from controlled experiments was 59-ft.

It has been stated by Dr. Gottwald that a key meeting was convened in December 1998. According to Dr. Gottwald, the consensus of attendees at the December meeting agreed to change the radius to 1900-ft. Since the 1900-ft rule was a massive departure from the prior protocol of 125-ft involving both federal and state agencies. it was unlikely to be decided upon at one meeting. More likely, concurrence came slowly from May to November 1999 at Task Force meetings. New legislation had to be enacted prior to the implementation of the 1900-ft rule. The decision makers, including Dr. Stephen Poe of the USDA/APHIS, Deputy Commissioner Craig Meyer and Commissioner Bob Crawford, would need to know the cost of eradication, manpower requirements for inspections and program time estimates, for planning purposes before approving the 1900-ft rule.

2. Lessons from Canker Wars Prior to Year 2000

The lessons learned should begin with the writings of Dr. Whiteside towards the end of Canker War II. He published three articles in which he explained how the prior eradication program (Canker War II) likely would not achieve its objective of complete eradication.[22,23,24] In these articles, Dr. Whiteside noted that there would always be areas in the groves or nurseries where incomplete inspections would result in isolated pockets of canker infected trees. This is in part because inspections could never find 100% of all canker infections.[24] Also, a return of citrus canker was possible through importation of diseased seedlings into the nursery stock of Florida. Remnants and reintroduction would thwart eradication efforts.

How much guidance could authorities gain from the prior eradication effort? Eradication in Canker War II was concentrated in nurseries and groves. When canker was discovered in a nursery, this generally meant destroying the entire nursery, usually burning it down. The Department was sued by the owner of the Polk nursery because of destroying "false canker." Scientists in court were testifying against the Department, and urging the USDA to end the quarantine of false canker. This was hardly the program the Department wanted to repeat!

Probably, the most important lesson would be that it is possible to spend 200 million dollars and be back battling citrus canker again, just 21 months after declaring Florida to be canker-free.

The moratorium on healthy tree cutting began in February 26, 1998. The 125-ft removal policy had been in effect for only 13 months. This was likely insufficient time to tell if 125-ft policy was working. There was no real assessment of the 125-ft policy by either the USDA or FDACS.

3. Laying the Foundation for Large Radius Cutting

The acceptance of the 1900-ft rule proceeded in steps. The first step was to obtain a sufficient consensus among the various agencies associated with the CCEP to support in principle the use of a larger radius.

From year 2000 to 2006, the Department kept to its narrative that 1900-ft was based on scientific study. There were many variations on this theme, but most referred to an epidemiology study with 19,000 citrus trees. These various statements are summarized in Short Note 6.3. In addition, the Department insisted they were just following the advice of the USDA. This added credibility since the USDA, with approximately 100,000 employees, would not mislead Florida. However, the Department seemed at a loss to name one other USDA- ARS employee, other than Dr. Gottwald who was involved in the field study.

The 1900-ft policy had some terrible consequences. In a worse case, one incorrect positive identification could result in the destruction of 780 healthy residential trees (~3 citrus trees/acre[19], 260 acres) if there are no overlapping circles. Under the 125-ft policy, one incorrect positive identification would results in approximately 3 - 4 trees being destroyed. However, the larger radius would mean less inspections and therefore fewer false positives.

Ken Bailey, Director of the CCEP in Miami-Dade and Broward counties worried about the safety of the inspectors if the radius was expanded to 1900-ft. as there could be a really backlash from residents. He stated at the Citrus Canker Task Force Meeting in May 14, 1999:

> Ken said he has people that have guns pulled on them every single day. You can call it scientifically right if you want to, but from a public relations standpoint, you are not just going to walk on these properties and start cutting 1,900 feet.

In the 2002 article on the field study, two problems plaguing the CCEP were identified.[9] First, the survey crews are unlikely to detect all disease in an area, because they rarely have 100% access to all properties. Secondly, the expanding quarantine areas made it difficult to resurvey all areas in a timely manner. The Department could not set up offices, hire and train surveyors as rapidly as needed. In the article, it was suggested that these problems and other program deficiencies, supported the 1900-ft policy. These comments were made more than two years after the 1900-ft program was implemented, but they likely reflect the thinking prior to year 2000.

The elusive nature of citrus canker disease is not unique. With other diseases, it is also likely difficult to define a specific area where new incidences of disease might appear. In control of these diseases, it is sometimes preferred to take control actions in an area far beyond the likely limits of dissemination of a pathogen, simply as a precautionary action. The analogy to an escaped prisoner or the one spore which got away, might be appropriate. One may call this a safety factor, but if the treatment is possibly harmful, such as occurs with many pesticides, then this might be characterize as an excessive "harm factor."

A safety/harm factor was used in the 125-ft rule, as the most distant identification of canker bacterial cells was at 32 m (105-ft) from the source trees. In this case using an eradication radius was set at 20% higher than 105-ft, creating a circle approximately 40% larger in area, since the area was increased from 0.8 to 1.12 acres. With governmental actions where excessive application could result in harm, the regulatory action should be within the "narrowest limits of necessity" consistent with the application of strict scrutiny, not what might be most efficient or as a remedy against uncertainty or limited resources.

At the International Citrus Canker Research Workshop in June 2000, there was a general discussion of eradication policies and inadequacy of a single inspection to detect all incidences of canker in groves. According to the transcript of the Workshop, on page 319, Dr. Gottwald suggests that a "fudge factor" is needed in applying eradication zones, to insure complete capture of subsequent infections, the equivalent to a safety/harm factor.[25] Thus, it seems that his solution to the detection problem and a need of multiple inspections is to eradicate more trees.

Dr. Gottwald also states that with more research, the eradication policy could be closely tied to the anticipated dissemination. This would likely be limited to eradications in groves, where risk assessments were conducted and variances to the policy were recommended. In June 2000, the eradication policy had advanced to massive cutting program within residential areas of Miami-Dade and Broward counties.

Ken Bailey at the June 30, 1999 subcommittee meeting stated the reason 125-ft policy failed was because they did not have the resources to remove the trees quickly and that resulted in subsequent infections. As reviewed in Chapter 1, Deputy Commissioner Meyer supported the 1900-ft rule as being more effective, as he expressed in the November 16, 1999 meeting.

> Once we get the cutting done in that area, it is three quarters of a mile area we never have to look at again except periodic annual drive-by and pick up a citrus tree if replanted and the sprout issue we will need to deal with. By extending the eradication area, we will reduce the survey need.

Note that there may have been a typo in the minutes, as the 1900-ft rule removes 40% of a square mile or 260 acres.

It is believed officials were concerned the risk of future outbreaks caused by "remnants" with the 125-ft radius. With an increased radius was there would be far less chance of leaving infected trees in isolated pockets once the program ended. The inspectors might have trouble returning to the properties repeatedly, but the cutting crews would only have to enter the property once. Yet, in pursuit of a rapid end of an eradication program, more trees are destroyed in residential areas than necessary.

To those who advocated a larger radius, it was likely the discovery of citrus canker in Manatee County on the west coast, gave extra impetus to increase the eradication radius. As Dr. Stephen Poe of the USDA/APHIS wrote in 1999 in the Environmental Impact Assessment, Florida officials did not inspect sufficiently the "high risk" areas in the 1992-1994 period, prior to a declaration of eradication. [19]

The situation would not be the same for eradication in groves or nurseries which would demand payments for their losses. Citrus canker was discovered in the lime groves in Homestead, but these groves were relatively small (~ 2800 acres or 4.4 square miles). Also, payments to grove owners would be the responsibility of the USDA.

4. The Miseducation Program (1992 - 2000)

Speculative or exaggerated claims on the destructiveness of citrus canker have a long been a part of the canker war dating back to 1915. This information would be comments made by the Department and published in the local newspapers. Some trade journals and agriculture extension websites also repeated this misinformation.

To justify the destruction of millions of healthy trees and the expenditure of millions of taxpayer dollars to what appeared to be a minor foliar disease, perhaps it is a prerequisite to portray canker eradication as a life or death situation for the citrus industry. It can be argued that this is another case of politics as usual. The canker eradication program was competing for state funds with other projects which must be funded by the state legislature and approved by the governor. However, the intensity of the miseducation program at so many levels and in different fields, ranging from economics to epidemiology, distinguishes this from politics as usual. The miseducation occurred at a high levels, including plant pathologists expert in the citrus canker disease from University of Florida's extension institute- IFAS, USDA/ARS and FDACS/DPI.

Some of the mechanisms of canker dissemination was described by scientists in a misleading and biased manner, as discussed in Chapter 3. The patterns of spread of citrus canker were based on sparse new discoveries of canker in areas which could not possibly have been thoroughly inspected and related to other discoveries miles away.[5,6] The highly speculative presentations were made by Dr. Gottwald, as the principal investigator joined with Dr. Graham, a researcher from the University of Florida/IFAS.[5,6]

The "new epidemiology" would allow that any new discovery in any part of Florida could be associated with any other discovery in Florida, to determine a hypothetical "distance of spread." The new epidemiology ignored the incomplete inspections in residential areas and the elusive nature of citrus canker. The difficulties in making a positive visual inspections of trees as high as 25 feet from the ground were reviewed in Chapter 1.

Drs. Gottwald and Graham presented a paper at the annual meeting of the American Phytopathological Society (APS) in August 9 to 13, 1997 in Rochester, NY , entitled "Analysis of the dynamics of spread of citrus canker in urban Miami" where canker had spread from 13 to 88 square miles (56,328 acres) in 17 months. This would translate in approximately 200,000 residential lots, as a common lot size is a quarter acre. The presentation abstract states, "The age of the infections in newly infected areas corresponded well with individual tropical storm events that contributed to subsequent spread of *Xac* ca. 15 mi. to the northeast." Abstracts have been posted on the supporting documents website.

The Miracle Raindrop

Department officials had a difficult time convincing the general public that citrus canker could be disseminated by storm event up to a distance of 1900-ft 95% of the time and beyond 1900-ft, approximately 5% of the time. The maximum inter-tree distance given in the 2002 published article was 2.2 miles.[9]

Consider the journey of the "miracle" raindrop, which defies the law of gravity, wind shadows, effect of UV rays and physical impediments which limit travel and is able to transmit the disease long distances as given in the 2002 published article.[9] As the raindrop passes through the canopy of a citrus tree, it comes in contact with a canker lesion for a sufficient duration to allow bacteria to ooze out of the lesion. The droplet coalesces with other droplets on the stems, leaves and fruits of the tree. Then it continues its journey, through the baffles of the canopy and is then transported by wind over fences, houses, cars, roads, parking lots and numerous other obstacles before entering the canopy of another citrus tree with sufficient concentration (generally greater than 10^4 CFU/ml is needed), on susceptible leaves (new flushes high in the tree) to initiate a new infection. To travel 1900-ft, the rain droplet would need to pass through 10 to 15 yards in a residential area, assuming ¼ acre lots. For the other distances within the 2002 article, rain would need to cross highway I-95 with 8 lanes of traffic each way, lakes, schools, and commercial centers.

The longer the distance, the more implausible this mode of dissemination becomes. There is no evidence that a rain droplet could pass through the canopy of citrus tree and be able to be blown over a single house or across one road. Articles published by Drs. Gottwald (USDA/ARS), Schubert (FDACS/DPI) and Graham (UF/IFAS) emphasize long distance movements of citrus canker by garbage collection, water meter readers, lawn maintenance crews, insects and birds.[5,6] Also, the opinion was offered in these articles that citrus canker could inadvertently be transmitted by handling citrus fruit that had been exposed to canker, although no symptoms were apparent on the fruit.

These speculative pathways and exaggerated distances were used to legally support entry into residential backyards without permission. Further, it would support the contention that healthy trees could not be grown in the groves, unless canker was eliminated in residential areas. No other country was eradicating canker in residential areas, except at the periphery of groves, to prevent dissemination to commercial settings.

An observational study can calculate distances between any two infected trees, perhaps miles apart. While the distance may be correct, the interpretation of this distance can be meaningless if attributed to windblown rain. This idea is stated more formally in Schubert et al. 2001 article in Plant Disease, where "outposts of infection" could be two infected trees or collection of infected trees separated by a long distance:

> However, the fact that intervening uninfected citrus exists between these outposts of infection makes the case seem stronger for human involvement *[than weather events]* in most long distance inoculum dispersal. This assumes that a disease gradient of some form should exist between source and destination of inoculum distributed by weather events, but this may not be a safe assumption. It is frankly impossible at this time to be absolutely certain how long-distance movement of inoculum occurs.[18]

This is why, experimental studies are critical to a realistic determination of the maximum distance of dissemination.

5. The Moratorium (Feb 26, 1998 to June 17, 1999)

During the moratorium period, the Department removed only infected trees in Miami-Dade and Broward, according to Commissioner Crawford's press release of February 26, 1998. It may have seemed like the moratorium ran counter to everything the Department had said up to this point, that canker could spread like wildfire and only cutting both healthy and infected trees could stop it, so the industry could survive. In fact, in year 2001 in the Miami-Dade court case, an attorney for the Department would use the wildfire analogy to explain why they should be exempt from search warrants.

In his February 26, 1998 press release, Commissioner Crawford stated that he was working with the Florida legislature as well as both public and private agencies to fund a reforestation program designed to replace removed trees with healthy, new citrus trees. The release also states that the cutting of infected trees only would accelerate eradication program. A one-year research project is established "to track the spread of citrus canker from infected trees to exposed ones."

True to his word, Commissioner Crawford announced on November 3, 2000, a 7 million dollar agriculture emergency fund to purchase citrus trees for all South Florida property owners who had lost their trees to the disease. No citrus trees were ever purchased for homeowners.

In the published article on the field study, Dr. Gottwald indicated that during the study period, infected citrus trees within the Miami-Dade study sites were not removed.[9] However, infected citrus trees in the Broward sites were removed.[9] The impact of removing infected trees is discussed in more detail in Appendix B.

Dr. Graham, professor of plant pathology at University of Florida, wrote in 1998:

> The eradication agencies are even more hampered than ever by groups of residential property owners who have legally impeded access to their property for survey and successfully lobbied for a moratorium on destruction of exposed trees.[11]

Yet, when they decide to restart the 125-ft eradication program in June 1999, a town meeting was held in Miami Springs, a small municipality in Miami-Dade County, to air any grievances, the Miami Herald reports that only 10 residents showed up. Miami-Dade County has a population of 2.3 million residents, so there was an outpouring of apathy rather than concern.

In 1998, the Department still owed grove owners 17 million dollars for wrongfully destroying what they thought was citrus canker during Canker War II. So, there likely was a distrust among grove owners of the Department. The Department's refusal to compensation to owners was based on a hard fought battle during Canker War II. The courts ruled then that exposed trees had no value, but the ruling was applied to commercial trees under the 125-ft rule. In 1998, there was an active class action lawsuit against the Department (*Valera* case) for cutting exposed trees using the 125-ft radius which would inevitably challenge the right of healthy tree cutting in residential area.

Funding Issues

The citrus canker eradication program likely was underfunded in the initial three years due to the expensive and controversial Medfly eradication program. A total of 32 million dollars was spent on aerial spraying from Tampa to Orlando.[14] The CDC estimated that 132,000 people were exposed to Malathion insecticide and in 123 cases, were probably or possibly made sick from the spraying. [1] The Medfly attacks many commercial crops in Florida, including strawberry, tomatoes, mangos, avocado and citrus, so it is considered a great threat to Florida's agriculture. By December 1998, FDACS had declared victory on the Medfly, likely giving some priority to CCEP funding.

Governor Lawton Chiles had asked for only 4 million dollars in the fiscal year 1998 to finance the ongoing eradication program.[13] The Governor had tried unsuccessful to tap into a grower's emergency fund, which was part of the "box tax" program to pay for the CCEP.[14] This was opposed by Commissioner Crawford and the citrus industry. On December 12, 1998, Governor Chiles suffered a heart attack and died at the Florida Governor's Mansion, leaving Lieutenant Governor Buddy MacKay to serve the remaining 23 days of Chiles' unexpired term. Jeb Bush became the governor in January 1999, and likely was more receptive to the eradication of canker using state funds.

There was some support for the program from the federal sources. In April 1997, the canker fight received an additional boost due to a settlement of 17 million dollars with the federal government, on a portion of the 70 million dollars cost which the state incurred during Canker War II. [15] In March 1999, an additional 25 million dollars was made available from a federal program against invasive plants and animals to fight citrus canker. The funds was split with the CCEP receiving 20.7 million dollars and the remaining 4.3 million went to pay for border inspection and quarantine for imported fruit. Vice President Al Gore made the announcement. [13]

The Naples Daily News reported that citrus officials had been in Washington just two days before the announcement, lobbying members of congress for this funding. Doug Bournique is quoted in the article as follows: "They ran out of money a long time ago and it was getting desperate. This gives them the money to really get ahead of this disease and not behind it."[13] So, a very obvious reason for curtailing the cutting of healthy trees for 16 months was the lack of funds. It is likely that funds were approved more easily after the Medfly crisis passed.

The One Year Time Table

At the Broward Court trial in year 2000, a manuscript of an article authored by Dr. Gottwald and four other plant pathologists was submitted to the court. The article discusses the negative reaction of grove owners and residents due to inadequate compensation with the 1900-ft policy. The article then states:

> Despite these problems, the political pressure exists for the eradication effort to go ahead with the ambitious goal of removal of all diseased and exposed citrus from Miami outbreak area within one year.

The sentence does not state that the program would end within one year, but canker would be gone from the "Miami outbreak area" by the end of year 2000. This sentence was not included in the article as published in January 2001. A copy of this manuscript is provided on the website. The decision to implement the 1900-ft likely was influenced by unrealistic goals to complete eradication in Miami-Dade and Broward counties by the end of year 2000. Cutting more meant no re-inspections were needed. Immediately following Judge Fleet's decision, public relations officers with the Department were stating that the decision had stopped the five year program just five weeks before its completion.

6. New History of the 1900-ft Eradication Policy

In Chapter 7, it was concluded that the field study was not the basis for the 1900-ft policy. This conclusion is consistent with Dr. Gottwald's various statements that the policy did not evolve from the October 13, 1999 interim report or any other report. His testimony at the Broward Court trial in November 2000 was as follows:

> No, the report *[1999 Interim Report]* does not discuss the meeting that occurred in December 1998 in which there was a group of scientists, regulatory agents and growers present in a room and decided upon 1900 feet. 1900 foot *[rule]* is not decided upon this, the manuscript or anything else, it was decided upon by a group, a body of regulators, university scientists as well as ARS scientists, etc. It was not decided upon by these reports.

The words in brackets were added for clarity. It is considered that part of the above testimony is correct; the 1900 ft rule was not decided by any report. The other part is likely only partially true as it is believed the December 1998 meeting was not really the critical decision meeting. In Chapter 8, evidence was presented suggesting the real origin of the policy came from a clear-cutting simulation model. This type of model is commonly referred to as Monte Carlo or random number simulation. The simulation model provided the percentage of areal coverage for 1900-ft and other radii including 125, 1200, 1600 and 2700-ft. It is also assumed that for the radii of 125 and 1200-ft, the simulation model would include estimates of a second inspection, which would result in more infected trees, and a second eradication effort.

By the time the Task Force first convened in early 1999, the Department limited the choices to the representatives of grove owner and packinghouses— either stay with the current 125-ft, which Dr. Gottwald would state would never catch up with the epidemic or the 1900-ft rule which could ultimately succeed. The main concern at the time, was that the canker would not move northward from Miami-Dade and Broward counties and threaten commercial groves in Central Florida.

Dr. Gottwald ultimately published two articles, suggesting that the 125-ft policy was insufficient to capture the subsequent occurrences of the disease and that the "distances of spread" were in many cases, far greater than 125-ft. It is likely this was sufficient scientific evidence for the Deputy Commission, whose chief concern was a legal challenge in the future. Deputy Commissioner Craig Meyer stated at the July 16, 1999 Task Force meeting:

> The scientific paper with the 1900 feet data will eventually be tested in court and as to how much farther beyond 125 feet can we take out without having to pay compensation to the owners if the owners are resistant.

Craig Meyer was not referring to any existing scientific paper, nor at this point any manuscript awaiting for publication. At this point, he was speaking hypothetically. The Deputy Commissioner likely knew that Dr. Gottwald would be making a poster presentation at the 1999 APS Annual Conference, held on August 7-11, 1999 in Montreal, Canada. The only public information from this meeting would be a brief abstract, not a scientific paper. However, the abstract would contain the values of 1200, 1600 and 1900-ft, corresponding to 90, 95, and 99% probability levels.

The "court test" was envisioned as something that could happen after cutting down the citrus trees. Later in the July 1999 meeting, the Deputy Commissioner Meyer states:

> We will be sued and are going to be asked in court as to why we go to 1900 feet instead of 125 feet and the answer will be because of this group's meetings, because of Dr. Gottwald's research and research done by others and because field experience shows that 125 feet in some cases did not work. All those will be in an argument held after the trees are removed are seeking financial compensation for those trees and this will be an important part of why the Department has moved beyond the 125 feet.

Craig Meyer seems to be holding four strong cards to defend the 1900-ft policy—Task Force recommendation, Gottwald's research, collaborating studies, field experience— in his imagined legal defense of the policy to the Task Force, but it would all fall apart in November 2000. The Task Force recommendations would not be considered important by the court, because this group was hand picked by the Department with no legal authority.

Solid Peer Reviewed Science

The "research done by others" or collaborating studies, as Mr. Meyer felt necessary to defend the 1900-ft rule in court was non-existent. While many other articles had commented on the viral nature of canker, none had suggested a large cutting radius. The potential for long distance movement of canker by windblown rain from observational studies came principally from Drs. Gottwald and Graham.

The Department in a letter to President Clinton in year 2000, indicates other epidemiologists supported the "1900-ft study."

> Concerning the validity of Dr. Gottwald's et al., citrus canker epidemiology research, work that is in prepublication form, it is important to point out that it has been reviewed by several of the world's leading plant disease epidemiologists. Dr. Gareth Hughes, of the University of Edinburg, Scotland, and a leading expert, has working closely with Dr. Gottwald in the analyses of his data and if anything, feels it may be conservative in its conclusion.

Dr. Hughes was a co-author to the brief, five page Letter to the Editor as published in Phytopathology in January 2000. He never publically made any comment on the research, and he is not one of the co-authors of the more extensive article on the field study in April 2002. He was never called to testify for the Department.

The Inner Circle

It is suggested that the move to 1900-ft was not based on an "epidemiology research" reviewed by several of the world's leading epidemiologists, but decided upon by a small group of plant pathologists. This inner circle was composed of three non-Department officials/researchers: Dr. Stephen Poe, USDA/APHIS, Dr. James Graham UF/IFAS and Dr. Timothy Gottwald, USDA/ARS. In addition, the inner circle included Mr. Richard Gaskalla, Mr. Leon Hebb, Dr. Timothy Schubert and Dr. Wayne Schubert, all of whom worked for the Department. This would make Dr. Graham the only researcher, not affiliated with neither the USDA nor FDACS. The inner circle, with the exception of Mr. Gaskalla, met numerous times to make recommendations on risk assessment, mainly for making allowances for commercial grove owners under the 125-ft rule.

Dr. Gottwald had been the author or co-author on two articles published in trade journals prior to January 2000, suggesting long distance dissemination based on windblown rain. The necessity of massive tree cutting based on field experience was used in court by calling Department scientists and Dr. Gottwald to testify, but this testimony was countered by the opinions of Drs. Whiteside and Wulscher. Thus, the Deputy Commissioner depended nearly entirely on one epidemiologist, Dr. Gottwald, for making the case that field study supported the 1900-ft. Dr. Gottwald may be considered the star epidemiologist for the program, but also its only epidemiologist.

7. December 1998 Meeting

One of the strangest parts of the Department's narrative is the mysterious December 1998 meeting. If one assumes that the field study began in August 1998, corresponding to the date of a joint agreement between USDA, FDACS and UF/IFAS, then December 1998 would mark the fourth month of a year long study. It just seemed odd for Dr. Gottwald to be providing preliminary field results at this point. Lesions take time to form and be detectable, so the elapsed time would hardly seem adequate time to obtain meaningful set of results.

According to the Broward Court transcript, both Dr. Gottwald and Deputy Commissioner Craig Meyer testified in court that there was meeting at the USDA/ARS office which was in Orlando, Florida in December 1998. Deputy Commissioner Craig Meyer testified this was an ad hoc meeting, arranged by Dr. Gottwald of the USDA/ARS, so he did not know who was invited. Mr. Craig Meyer testified that he was present at the meeting.

Dr. Gottwald testified in the Broward Court, concerning the December 1998 meeting:

> At that meeting, this data was poured over by a set of regulatory individuals, university scientists, and leaders of the citrus industry, who all examined this data. A consensus was- we came to a consensus as group, so this is not myself saying this, that we're using 1,900 feet, but the consensus of a body of individuals familiar with the disease and familiar with the status now.

The Department could not offered any tangible piece of evidence that the December 1998 meeting occurred. There are no list of attendees, no minutes, no copies of presentations or notes made at the meeting Although several comprehensive articles were published from 2002 to 2004, specifically detailing the "epidemiology research" lead to 1900-ft rule, not a single article has ever identified a specific date in December 1998 in which this meeting supposedly took place. Did it take place at all?

According to the Department, the ad hoc meeting was well attended. Representatives of the citrus industry were present according to court testimony. Dr. Gottwald testified he could not remember if the meeting was noticed and advertised. He testified he believed the press was invited. A thorough search through all news media articles written on citrus canker showed no mentioned of the meeting in December 1998. Judge J. Leonard Fleet was disturbed on how the general public had been ignored as he wrote in his November 17, 2000 opinion:

> The attendance at this meeting, whether by design or accident, had no representatives of the general public in attendance although scientists and delegates from the commercial citrus industry were invited to attend. The extreme importance of this meeting is underscored by the undeniable and admitted fact the Department implemented the precise recommendations contained in Dr. Gottwald's presentation.

Judge Fleet cites more examples where the public was systematically excluded from the decision making process, and concludes:

> President Harry Truman was right when he said secrecy and a free democratic government don't mix.

Alternative Interpretation - A meeting of the inner circle of colleagues

There is no way to confirm the December 1998 meeting took place, let alone know who was there and what was presented. The meeting might have been of "extreme importance" as Judge Fleet opined, but not for the reasons he gave. The pertinent question should be why would Dr. Gottwald want to attach such importance to a meeting in which there were no minutes, no list of attendees, or no prior notices. After all, the 1900-ft policy had been discussed at three Task Force meetings in 1999, with detailed minutes. It seemed strange that there was no mention of a December 1998 meeting among key stakeholders in any minutes of the Task Force or reports from the Risk Assessment Group.

However, if the meeting did take place, it could have been a regular meeting of the Risk Assessment Group, with Mr. Richard Gaskalla and Craig Meyer present as invited guests, with a total of 9 attendees. In this case, Dr. Wayne Dixon of the FDACS/DPI would call the meeting as he was in charge of group. It is possible that in December 1998 had nothing to do with the field study, but everything to do with the simulation modeling. Dr. Gottwald interest in 1998 would be to present the simulation work just to an inner circle of colleagues, and key decision makers, in particular Dr. Poe of the USDA/APHIS and Deputy Commissioner Meyer.

In a sense, the simulation work was a clean set of results as it did not require observational data. If 1900-ft circles were drawn around any 10 objects randomly located in a square, then these circles would, on the average, cover 95% of the area. Of course, it had nothing to do with the distance of canker transmission. Moving the decision date back to December 1998, clearly avoids the May 1999 meetings and the July 16, 1999 Task Force meetings, where field study results were presented.

8. Presentations and Decisions in 1999

May 1999 Meetings:

On May 11, 1999, the RAG meeting was held, with all seven members in attendance. In contrast to the December 1998 meeting, this meeting was documented by a short report. At the meeting, Dr. Gottwald presented the field study results. The meeting's report provides very few details of the study. Two motions were presented to the group. The first was that the exposed trees should be removed to within 1900-ft of a positive tree. The second motion recommended removal of trees first in Broward County and then proceeding in a southerly manner to continue removals in Miami-Dade County. It recommends that trees be initially removed to 125-ft, but as resources are available, the eradication circle would be expanded to 1900-ft. The members vote unanimously to approve both motions.

Then three days later, on May 14, 1999, the Task Force met with 51 attendees. Dr. Gottwald made a similar presentation, and once again, there were few details in the minutes of what has been presented, except for the identical table of results from the prior presentation. However, after the presentation, there was no action requested of the Task Force. It appears the Task Force was unaware of the May 11, 1999 recommendations until the October 19, 1999 meeting.

In the figure below, the Site 1 results as presented by Dr. Gottwald has been reproduced. In the third period, the values 1200, 1600 and 1900-ft correspond to the capture levels of 90, 95 and 99% of the subsequent infected trees. In the Task Force minutes, the value of 1900-ft is highlighted. It is alleged that in place of field observations, simulation results were substituted in the third period of the values for the one month windows, thus the results were a mix between field results and simulation work, but presented as if they were field results.

Thus, a concocted "sandwich" presentation of results was created, so 1900-ft was neither the highest or lowest distance necessary to capture, but something in between these values. This would be good for public relations, they would contend that a much higher capture radius could have been used. It would also be good for legal reasons as discussed in the next section.

Table 9.1: The Sandwich Presentation of Results

Site 1 values presented at the May 11, 1999 CCRAG Meeting with Period 3 taken from the simulation model results

	# of focal trees	# Newly Infected	90% Capture	95% Capture	99% Capture
1st Period	38	15	800	4150	4150
2nd Period	52	38	1450	1650	1650
3rd Period	**90**	**73**	**1200**	**1600**	**1900***
4th Period	162	235	700	800	1450
5th Period	396	124	350	300	700
6th Period	519	32	250	950	950

* Third period distance values allegedly taken from simulation model runs, and not from field observations, resulting in a dishonest presentation of results.

In the minutes of the meeting, there is a footnote at the bottom, which reads, "Updated May 26, 1999", so technically it is unknown what was presented by Dr. Gottwald at the meeting. However, these results are the same as presented in the October 13, 1999 interim report and the November 2000 Broward Court hearing. The latter presentations included Sites 2, 3 and 4.

June to November 1999 Task Force Meetings

In Chapter 1, a few excerpts from these meetings were provided. These three meetings were key to the final implementation of the 1900-ft rule.

June 30, 1999: Joint meeting of the Regulatory Issues and Science Issues subgroup of the Task Force. Members approved a motion recommending the 1900-ft with the added words "based on risk assessment." The risk assessment policy was not well identified, but it is likely the citrus industry associations felt that the Department would not eradicate without sufficient compensation. Resource availability was one of the factors in the risk assessment.

July 16, 1999: General Task Force Meeting. Members approved the 1900-ft policy conditional on risk assessment. There was considerable discussion if the wording meant that risk assessment could allow grove owners to protest a pending eradication action, or it would be used by the Department prior to an eradication action. The discussion seems to indicate risk assessment would be used both ways. The minutes are not clear on what recourse the grove owners would have if they disagreed with the Department's risk assessment. Mr. Gaskalla states that in the prior eradication program, this was left up to the Technical Advisory Committee. It is likely that following the July 16, 1999 meeting, there were more discussion on the funding and logistics of the program.

November 16, 1999: General Task Force Meeting. This is the final Task Force Meeting prior to the implementation of the 1900-ft rule. In Dr. Gottwald's 2001 article, he states the reasons for the delay in the implementation of the 1900-ft rule:

> The timely implementation of the 1,900-ft rule and the sentinel tree grid system would not have been feasible in 1999, given existing funding and manpower constraints. However, once the results of the epidemiology study became widely known and understood, citrus industry groups pressured the CCEP for a more effective eradication effort. These groups immediately lobbied state and federal sources for stepped up financial support of the 1,900-ft rule. Within a matter of days, the Governor and Commissioner of Agriculture of Florida announced that 175 million of combined state and federal assistance.[8]

It is believed that Dr. Gottwald is referring to the November 16, 1999 Task Force meeting when he states "the epidemiology study became widely known and understood" to citrus industry leaders. Both Dr. Gottwald and Mr. Richard Gaskalla urged support of the 1900-ft rule and following the November 16, 1999 Task Force meeting, it is likely the citrus industry groups lobbied for approval of compensation and an industry-friendly risk assessment program to go along with the 1900-ft rule.

The "results of the epidemiology study" is likely a presentation made by Dr. Gottwald on the potential of tropical storms, hurricanes and tornados to disseminate citrus canker from the south of Florida to the central growing regions. The November 1999 presentation warned growers — citrus canker is moving north and you do not have much time left.

The minutes of the meeting state, "Tim *[Dr. Gottwald]* talked about the different hurricanes that have occurred this season (i.e., Floyd, Harvey and Irene). Hurricane Floyd skirted up the coast and Tim reported that he wanted to use that as an example of what was going on with a hurricane..." Further in the meeting, the minutes state:

> Tim *[Dr. Gottwald]* said he is going to make a plea and it is that going to 125 feet isn't going to do it, nor will 800 feet. If you want to have an effect, you will have to take much more out. Normal rain storm events can spread the disease 1900-ft.

Richard Gaskalla states later in the meeting:

> ...if you go into a quarter section and remove all the trees, that is a quarter section that you don't have to go back into.

The area of a quarter section is 160 acres, but from the context of the discussion, it is clear Mr. Gaskalla was discussing the need for a 1900-ft radius, in line with the general discussion of reducing the number of resurveys.

The situation may not have been exactly as Dr. Gottwald had stated with respect to the citrus industry pressure officials for a change in policy. Governor Jeb Bush had just been elected, and most likely was reluctant to make a change in policy, unless compensation had been approved and funds made available to grove owners. Sometimes, approval of funds, and their availability do not occur simultaneously.

Strict Scrutiny and Compensation

It is easy from a legal perspective to know what Deputy Commissioner Meyer did not want — the court interference with the eradication program. From Mr. Meyer perspective, all the Department was trying to do was to rid Florida of a menacing disease, and if they were going to be checked every step of the way, by a meddlesome judge, then their program was never going to succeed.

The Deputy Commissioner Meyer by training was a lawyer, and understood well the legal issues with the program. The Corneal case decided in 1957, opined, "The destruction of private property must be within the narrowest limits of actual necessity unless the state choses to pay compensation." This is referred to as strict scrutiny.

Providing compensation for the destruction of exposed trees ran counter to the hard fought court battles, in which the Department insisted that exposed trees had no value. They would eventually show signs of citrus canker. Then, in late 2001, the Department suddenly seem to change course, with the passage of Florida Statute 581.1845 providing compensation to homeowners. What had changed was the Broward Case had now entered the Administrative court, and the challenge under strict scrutiny was a real possibility because the redeemable Walmart vouchers for garden supplies given to homeowners were never considered compensation.

The canker compensation law was enacted to circumvent a trial in the Administrative Court. The strict scrutiny standard meant the Department would have to show the 1900-ft was truly a policy "within the narrowest limits of actual necessity." If the plaintiffs were able to prove that the 1900-ft policy originated with a simulation model, then the concept of the exposed trees being the next victims of canker would have been proven false. The simulation model simply provided estimates of areal coverage, or how overlapping circles would cover an area. There was no biological association between infected and healthy trees in the simulation.

9. No Easy Exit Plan

As discussed in Chapter 2, Canker War II ended with apparent success, because the Department did not make an all out effort to inspect the high risk areas. This comment came from Dr. Stephen Poe, who was part of the Task Force in the 1980's, directing the program.

The Department recognized it may run out of funds, and needed a way to cut back in the future. This is why the risk assessment had as one of its factors, resource availability. However, using risk assessment to reduce the cutting radii of commercial growers (grove and nursery owners) would have been a disaster. It would be obvious to residents that canker would be more rapidly disseminated by windblown rain within groves and nurseries than in residential areas. Once the 1900-ft policy had been implemented in early 2000 in the residential areas of Broward and Miami-Dade counties, and with over half a million residential trees and 50% of the lime groves destroyed, reducing the cutting radius for commercial interests would have greatly angered residents. It should be remembered that the Commissioner of FDACS is an elected official.

Also, there was no easy exit as done in Canker War II with scaled back inspections. It is likely this was how the program ended in Canker War I. The USDA sentinel program was designed to broaden the scope of the CCEP surveys and discover citrus canker throughout Florida. Although residential discoveries in Central Florida tended to be small, such as the one in Okeechobee County with one positive tree, the entire county would be colored red, as another canker infested county.

What really made the program inflexible, was the passage of the "new canker law" in March 2002. Designed specifically to circumvent the Administrative Court examination of the 1900-ft policy, it removed the Department's ability to implement new rules through risk assessment for a graceful exit to the CCEP. The new law proved to be a disaster, because the legal challenge would be back in Judge Fleet's court, and this time, the Department could not argue it belonged in Administrative Court.

However, what fueled the program at this point, was the Department's own rhetoric – canker would destroy the economy of Florida. The hype was tremendous. Plus, on all other fronts, things were positive. Dr. Gottwald had published the Letter to the Editor in January 2001, with the 11.1 mile "spread distance" and was working on a more comprehensive article, which would be published in April 2002. And compensation money was flowing to the South Florida lime grove owners, so they could take advantage of the housing boom.

Dr. Griffiths quip summed up the situation well, as he told an audience, "I suspect a lot of you would like to have canker" at the annual Citrus Growers Association (Kevin Bouffard article, Lakeland Ledger, Dec 5, 2002). As long as the USDA cash was flowing to compensate grove owners, the program would continue.

10. The Dissemination of Citrus Canker throughout Florida

In Chapter 6, it was concluded, based on a review of the technical literature, that controlled experiments have not demonstrated dispersal of citrus canker from an infected tree to a healthy one beyond a distance of 59-ft by windblown rain. These experiments were conducted under ideal conditions, with an unobstructed pathway between infected and uninfected trees citrus trees.

It was noted that the objective of these experiments was not specifically to identify the maximum distance which citrus canker bacteria could be dispersed and cause infection in host trees, under conditions of windblown rain. The experiments were not representatives of conditions within residential areas. In these areas, homes and other fixed structures would act as windbreaks, effectively impeding the dissemination of citrus canker.

But these experiments raise the question of how could citrus canker become so wide spread within Florida. In the Broward Court Case 2, Judge J. Leonard Fleet after hearing testimony for two weeks from experts in both epidemiology and plant pathology, was not satisfied with the explanation of canker dissemination:

> Based upon the evidence before the court, this explanation does not satisfactorily explain the appearance of citrus canker in diverse areas of the state separated by large expanses of geography.

Epidemics need both short and long legs to exist, to paraphrase a famous epidemiologist, Dr. Vanderplanck.[20] For canker, the short legs are typically from rain storms and irrigation systems. The nursery environment with the extremely high density of seedlings, presents the optimal conditions for spread to many plants.

A recent textbook on plant disease epidemiology[13] states, "Typically, diseased plants do not move. " So, it would appear citrus canker is not a typical disease. Epidemiology spatial models typically assume the diseased plants are in fixed locations and only the disease is capable of moving among plants within farms or orchards. With citrus canker, the disease can move with the nursery stock.

Prior eradication programs focused mainly on destruction of nursery plants. The first eradication program (Canker War I) in 1912 resulted in the destruction of 3 million citrus in the nurseries. It was concluded that the most likely source was imported citrus seedlings from Japan in 1910 in a shipment to Houston, Texas. Canker was discovered in six other states besides Florida and distributed to nurseries as far north as South Carolina. Canker War II was similar, as most destruction occurred nurseries.

It is suggested that the long legs of the Florida epidemic were contaminated nursery plants from Central Florida. Seedlings are likely well inspected, but if the plants leave the nursery with lesions at an undetectable level, then they are likely to cause dissemination throughout Florida. The new history of the epidemic considers the Hurricane Andrew which came ashore in the

Homestead area of Florida in August 1992 as the single most important event in the epidemic. As described by Schubert et al, in a 2001 article,

> Hurricane Andrew hit South Florida in 1992 about the time that the disease made its appearance, and some have speculated about the storm's possible role in the introduction. Several other exotic pests appeared in South Florida at about the same time: the Asian citrus leafminer, the brown citrus aphid and the Asian citrus psyllid.

The Asian citrus leafminer (CLM) was reviewed in Chapter 3. The presence of leafminer is very clear with the rolled up leaves, and the cellophane-like mines created by the CLM lavrae.

It is believed that during the replanting of the lime groves, some of the seedlings which came from the nurseries had minute canker lesions. Disease is easily disseminated when the seedlings are closely packed in containers and regularly watered.

Hurricane Andrew devastated many areas including Sweetwater, Florida, where many small commercial outlets were located. In many cases, it took months for many of these outlets to rebuild and build back inventory.

The Department has implied that citrus canker originated in Miami-Dade County and then spread to other parts of Florida. However, they did not specifically present a map showing northern movement, as shown in Figure 9.2. However, the Task Force considered a firewall north of Palm Beach County, which would be zone clear cut of residents' tree. Considerable attention was drawn to the fact that the Miami strain was first present in Miami-Dade County and later, it was inferred that tropical storms were responsible for long distance dissemination to other parts of Florida.

An alternative version of canker movement is shown in Figure 9.2 with citrus canker originating primarily in Central Florida, and then moving to all parts of Florida as nursery seedlings. It is likely oversimplistic. Due to the elusive nature of citrus canker, the discovery date should not be considered the initial date of infection. Citrus canker bacterium was not disseminated long distances by windblown rain, but rather by trucks distributing contaminated trees to the east and west coasts of Florida.

The Comprehensive Report states for Polk County for May 23, 2005:

> Citrus canker infestation was confirmed in a citrus nursery on Hamlin oranges in Frostproof at 32S28E25. Trees were recently moved from this block to groves in Polk, Highlands and Hardee counties as resets. Some of the resets are exhibiting symptoms of citrus canker. Surveys of all plants moved from the nursery in the past year were conducted.

Further, on May 25, 2005, the Comprehensive Report states, "Control action of Ben Hill Griffin nursery completed."

Nurseries supply both groves and commercial outlets with citrus trees. To follow up on all shipments from a primary nursery selling citrus in liners is a very laborious task. The shipment to commercial outlets may be all sold to customers by the time inspections occur.

In commercial outlets, since seedlings are typically clustered tightly together, and with perhaps 20 or more plants in a space of 100 ft^2, citrus canker can very quickly infect every plant. Contaminated nursery stock arriving at large retailers, is likely quickly hosed down on arrival and then sold to consumers.

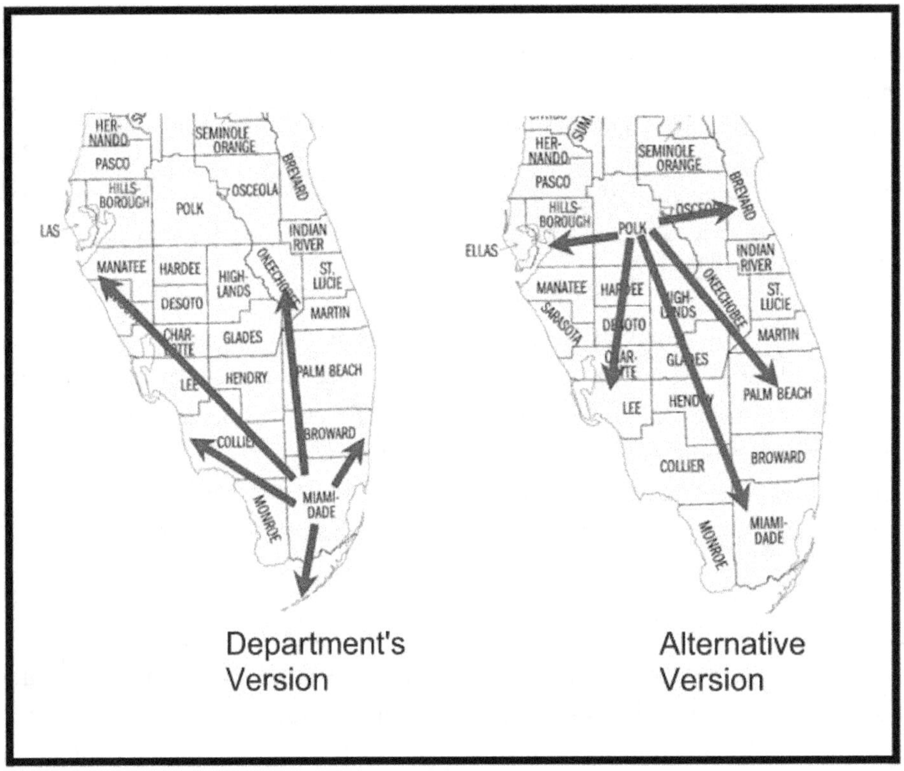

Figure 9.2: Spread of Citrus Canker throughout Florida

It is very possible that some of the discoveries came were remnants from Canker War II. Before Florida was declared canker free, there were no inspections on the east coast.

In summary, citrus canker was disseminated short distances by windblown rain. Long distance dissemination was the result of contaminated plants from nurseries in Central Florida being shipped throughout Florida. There were opportunities all along the supply chain for seedlings to become infected. No program completely eliminates citrus canker, so Florida has never been canker free.

11. The End to the Program

On January 10, 2006, Deputy Secretary of Agriculture, Chuck Conner announced the USDA was officially ending support of the citrus canker eradication program. It was likely no surprise to the Department. The future of the citrus canker program had been discussed in a closed door meeting between the USDA, FDACS and leaders of the citrus industry on December 12, 2005. The Deputy Commissioner stated there were numerous discussions on the future direction of the program in the prior year.

The USDA letter began with the conclusions that the series of 2004 to 2005 hurricanes had made the eradication more difficult, and if the program went forward, surveying and removals requirements would be much higher. Commissioner Bronson issued a press release on January 11, 2006 stating his agreement with the USDA letter and indicating the Department would work closely with the USDA with improved phytosanitation and disease control measures.

On balance, the Deputy Secretary's letter was quite positive to grove owners, as it committed the USDA to work with FDACS in a citrus health plan aimed at not only citrus canker but also citrus greening and other exotic diseases. The key elements of the Citrus Health Plan are summarized in Chapter 10.

Opponents of the program were likely elated with the news. Many had believed the CCEP was doomed from the very beginning. Opponents doubted there was solid evidence showing the 1900-ft was necessary. The Broward Court hearings had supported their views, with experts testifying that it was unlikely citrus canker would ever be eliminated. On January 11, 2006, Commissioner Bronson issued a press release and included legal delays as a contributing factor to the unsuccessful eradication.

The requirement of search warrants made inspections of residential areas much more difficult, but nurseries and groves could be inspected without search warrants. Through the long series of legal challenges, the Department was never barred from removing infected trees.

Hurricanes and Tropical Storms

USDA Deputy Commissioner Conner provided details on how canker had spread after Hurricane Wilma. Tropical storms are a common feature of summer months in Florida. No one could seriously fault the Department for not being able to adequately survey and remove trees in the groves after the four hurricanes passed through Florida. The last hurricane, Wilma, passed through Florida in a northwest direction on October 24, 2005. It was a highly destructive hurricane. Conner's letter mentions the Second International Citrus Canker Workshop convened in Orlando, Florida in November 2005, enabling consultations with a panel of global experts in citrus diseases. He states, "it was their conclusion that the disease is now so widely distributed that eradication is infeasible." A conclusion that a panel might caution officials on the prospects of a timely and full eradication is not surprising, given that the groves would be replanting after the hurricanes.

Conner's letter included a few statistics on CCEP costs and eradication status. Also, the letter included a preliminary projection of the consequences of Hurricane Wilma, based on the canker workshop in November 2005. It stated that the following Wilma, CCEP substantial progress was made on 32,000 acres of "affected" commercial citrus, but it was now estimating that 168,000 to 220,000 acres would also be necessary to destroy under the 1900-ft policy. Including the 32,000 acres, and assuming 130 citrus trees/acre, this is an astounding 26 to 29 million commercial trees. The letter does not specify how many of these would be infected trees. The total number of commercial grove trees destroyed since the program began was 11.3 million trees on 87,493 acres.

If the estimates were made in time for the workshop, the estimate had to be made in less than two weeks (October 24 to November 7, 2005). Thus, it is impossible for any canker incidences from the hurricane to be identified. The workshop included an abstract for paper C-5, entitled, "Post-hurricane analysis of citrus canker spread and progress towards the development of a predictive model for future weather related spread." Neither Hurricane Wilma nor the estimate of 168,000 to 220,000 acres being exposed is included in this abstract. All abstracts are posted online.

An article was published in 2007 by Dr. Gottwald and Michael Irey, in Phytopathology explaining the basis for the projection. It is based on a Monte-Carlo simulation model, which has many similarities to the clear-cutting model, as described in Chapter 8. In both models, many patterns or "mock epidemics" would need to be generated to obtain a statistically meaningful average estimate of areal coverage of the 1900-ft rule.

The hurricane model considered a single point source, and an exponential distribution (dispersal function) to identify locations of infected trees. The hurricane model used maximum dispersal distance from point source to furthest location of 30 km (98,400 ft or 19 miles). It was noted in an article in by Koizumi, that an earlier study had observed typhoons capable of dispersing citrus canker up to 15.7 m.[26] Thus, the distances presented by in the article are 2,000 times higher.

Undoubtedly, the 2004 to 2005 hurricane season increased the dissemination of citrus canker particularly in the groves. Winds in excess of 100 mph are highly destructive, as all Floridians are aware. However, as researchers have found through experimentation, heavy rainfall does not necessarily disseminate more bacteria than light rainfall. For any infection process, the rain must first pass through an infected tree, and there must be sufficient time for the bacteria to ooze from the lesions. The mechanics of a hurricane, with high gusts, breaking limbs from trees and reducing the canopy, would have a tendency to limit the exiting windblown rain. Prolong rainfall can results in diminishing quantities of inoculum from lesions, as later rainfall tends to wash off the bacteria. In the path of a hurricane, it is likely the damage would be intense, so grove owners might favor eradication programs to help clean up groves with generous compensation programs.

Observational studies can provide misleading results. It is likely difficult to distinguish between the effects of two hurricanes in one year, from the approximately 70 storms which occur each year in Florida. According to the Lakeland Ledger, when Hurricane Charley hit on August 13, 2004, the program had not ramped up to remove thousands of infected and exposed trees in Southeast Florida. So, it made sense that in 2005, there would be many more discoveries of citrus canker in groves, not necessarily caused by the hurricanes but just wet conditions. It is also likely that the nursery stock was contaminated with canker, thus as grove owners begun massive replanting after the hurricanes, some of these trees would start to show symptoms.

The hurricane study as published in Phytopathology in 2007 is not reviewed in this book, as the program had already ended. However, if FDACS ever wants to study the effects of canker dispersal during hurricane force winds, the Florida International University has the right tools, with the "Wall of Wind" hurricane simulator, capable of generating Category 5 winds.

Destruction of Nursery Stock

A factor causing the USDA to withdraw support was the depletion of nursery stock, due to eradication efforts in nurseries. Deputy Secretary Conner's letter discusses the depleted state of nursery stock, as follows:

> Still [*after compensation is provided*], it would take anywhere from 5 to 15 years for the industry to recover because the State's citrus nursery infrastructure has been seriously eroded by citrus canker, citrus greening and storm damage.

The Florida Comprehensive Report states that a total of 4.3 million trees nursery trees were destroyed by the end of the program in 2006. This is approximately 26% of all citrus trees cut in Florida and five times the total residential trees. The Comprehensive Report provides the names of five nurseries where citrus canker was discovered.

Table 9.2: Destruction of Nursery Stock

County	Date	Nursery	Trees destroyed**
De Soto	15-June-2005	Orange-Co, Arcadia	120,361
Hendry	27-May-2005	Duda and Sons	665,071
Highlands	23-May-2005	Unnamed *	1,871,346
Polk	25-May-2005	Ben Hill Griffin	1,677,375
Total			4,334,154

* The unnamed nursery is located at 33S37E36 in Venus, FL. It is an abandoned nursery with orange trees. ** Nursery trees destroyed include the specified nurseries, but may include other unnamed nurseries within the county in these totals.

The discovery and destruction of nursery stock in 2005 all occurred within approximately 22 days. Compensation varied depending on the size of the container size, ranging from $5.00 for a one gallon to $26.00 for containers above 7 gallons. How would anyone be able to check on the number of plants and containers in a nursery after the plants had been destroyed? The compensation program covered all certified stock dating back to 2003. It was reported that 65% of the nursery stock had been destroyed by the program.[17]

A nursery in a quarantine area essentially is out of business, because their plants can not be sold. However, a single discovery of citrus canker could be worth millions in compensation. The Palm Beach Post headline on December 6, 2005 read, "Canker Program has Citrus Nurseries Near Collapse." Chuck Reed, owner of Reed Brothers Nursery in Dundee is quoted as follows:[17]

> The private citrus industry is close to collapsing, in my opinion. I am in a quarantine area, which means I can't move anything in or out of my nursery for two years...You can't afford to maintain the trees for two years. We turned the water off and let the trees die.

Reed Brothers is a nursery that provides seedlings to other nurseries, so the supply chain to both groves and homeowners was in jeopardy. He stated in the article:

> I predict there will a time frame of several years that there will not be any nursery trees available for commercial groves. This will affect their bottom line drastically. Citrus for residential back yards will also be affected. We supply trees to nurseries that step them up in the bigger pots and sell them to homeowners.

The destruction of the 65% of the nursery meant that the recovery from the hurricanes was likely to be costly. Mr. Andy LaVigne of Florida Citrus Mutual had warned of this outcome in his opening remarks at the International Citrus Canker Research Workshop in June 2000:

> Many people have mentioned to me, well, where do you stop, you know, you get to a point where the citrus canker eradication program becomes a citrus eradication program. Well, exactly. We've got to decide that when we get to that point. Nobody knows what that point is. It is somewhere down the road.

The end of the program is blamed primarily on hurricanes and court injunction. The court injunctions were limited to cutting exposed trees in Broward and Miami-Dade County. The requirement for search warrants did not impact commercial groves and nurseries, as the State has a legal right to inspect and destroy plants without a warrant.

By 2005, the CCEP was looking more like CEP, or citrus eradication program, with over a billion spent and 16.5 million trees destroyed.

12. Concluding Remarks

A shaky foundation for the CCEP was built on a series of misrepresentations, exaggerations and occasionally outright lies. The biology was misrepresented by a fictional bacteria, which could hop rides on mail trucks, lawn mowers and meter readers and quickly spread throughout Florida. According to the Department, the destructive powers of the bacteria went to extremes, as the disease could wreck havoc on both residential and commercial groves, ultimately destroying the 100 million commercial trees in Florida.

The citrus canker was misidentified on many occasions, resulting in the unnecessary destruction of citrus in areas of up to 260 acres. Inspectors were not only misidentifying canker, they were occasionally misidentifying trees as citrus trees. Legally, there was little a homeowner could do. Homeowners were told this was all a necessary sacrifice to save the citrus industry.

The origin of the 1900-ft policy was misrepresented by FDACS. The well worn Department's version of the events recited by the Department goes something like this: "The 1900-ft policy was enacted after a 12 to 18 months long epidemiology study involving the monitoring of 19,000 citrus trees." Various versions of this story exists, but the general idea is that there was an extensive "epidemiology" study involved prior to the 1900-ft rule. so the Department is not removing too many or few trees.

Dr. Gottwald's Broward Court testimony that the 1900-ft rule did not come from any report is true. In fact, the new history is completely consistent with this part of his testimony. However, the 1900-ft policy was not the decision of unnamed group of persons in a room in Orlando, Florida in December 1998. The evidence points to an undisclosed simulation model as described in Chapter 8 as the real origins of the 1900-ft rule.

The statistical method of analysis or the "distance necessary to circumscribe" procedure has never been used since it was first published in year 2002. Scientists can not estimate the initial date of infection to track canker movements over long distances in residential areas. This is the reason it has failed the test of time.

Although nobody knows for sure, it is possible canker originate Central Florida. In this case, the primary sources were contaminated seedlings with lesions too small to be identified. It is suggested that canker easily worked its way through the supply chain, when there was a massive replanting in both groves and residential areas for several years after Hurricane Andrew.

Deputy Commissioner Craig Meyer may have had good intentions and in 1999 truly wanted a program which was well within the law and built on a solid foundation of technical studies. However, a study which would state unequivocally that the radius must be at distance *X*, based on scientific study to get rid of canker was not a realistic objective.

However, in truth, all Craig Meyer had in 1999 was the general alignment of four scientists (Dixon, Schubert, Graham and Gottwald), a couple of articles published in trade journals and a non-public report (interim report) from Dr. Gottwald, marked "Do not Distribute" followed by numerous exclamation points. It listed four other co-authors, which it is suspected never saw this report.

There were no collaborating controlled experimental studies showing long distance transmission of canker by windblown rain. Not in 1999, and even today, there are no published experimental controlled studies showing a 1900-ft or greater transmission of canker bacteria. This made any "bottoms-up" inquiry of the field study a real legal liability for the Department.

However, in implementing the 1900-ft rule the Deputy Commission settled for far less. By 1999, the Department narrowed the options between a 125-ft and 1900-ft policy. The efficiency of 1900-ft policy for residential areas had been emphasized by both Craig Meyer and Richard Gaskalla. It is believed that the USDA/APHIS was more concerned about a program that had a well defined timetable to a final declaration of a "canker-free" Florida.

It is believed the plan was to completed the eradication program before publication of the "epidemiology research." Thus, the necessity of the 1900-ft rule would be academic. As the field study was challenged in court, the USDA refused to release survey sheets and other basic information for inspection to the county attorneys All the county attorneys could do is to question Dr. Gottwald on what was done.

There were real legal issues concerned with basic constitutional rights of residents as guaranteed under the Bill of Rights against unreasonable searches, compensation for destroyed property and due process. Craig Meyer also seemed to trivialize the legal problems. If the legality of the program were challenged, Craig Meyer stated at the Task Force meeting, he could demand a bond equal to the value of the citrus industry. The Deputy Commissioner likely thought the eradication program might encounter resistance, but they would outrun any challengers.

At the foundation of the program was not good technical research or a solid legal basis, but incredible hubris and general callousness towards residents. How much of the legal and scientific problems with program permeated to the Commissioner of Agriculture, Bob Crawford, or the newly elected Governor, Jeb Bush, we will likely never know. Possibly, the Commissioner and Governor expected general support or mild opposition from the Broward and Miami-Dade County officials. It is expected no one in the state capital of Tallahassee quite realized the tenacity and persistence of the attorneys in the legal departments in Broward and Miami-Dade counties, nor opposition from County Commissioners and residents.

Bit by bit, the pieces of the puzzle have finally come together. This is the new history of the citrus canker program, fully supported by documents and reports from the USDA and FDACS. It has been hidden from the public until now.

Short Note on the Website:

SN 9.1 Still some missing pieces

10. Post CCEP: Living with Canker

What's past is prologue

William Shakespeare's The Tempest

The great enemy of truth is very often not the lie— deliberate, contrived and dishonest, but the myth — persistent, persuasive and unrealistic. Too often we hold fast to the clichés of our forebearers. We subject all facts to a prefabricated set of interpretations. We enjoy the comfort of opinion without the discomfort of thought.

John Kennedy, June 11, 1962, Yale Commencement Address

---

1. A New Day — Living with Canker

January 10, 2006, marked a turning point in the battle against citrus canker. On this date, the USDA withdrew financial support for the CCEP. Without the USDA's financial support, FDACS officially shut down the eradication program the next day. From January 10, 2006 forward, commercial grove owners would have to learn to live with canker.

No other country in the world would think of destroying 16.5 million citrus trees, and spending 1.3 billion dollars to eradicate citrus canker. It is unlikely, in 1995 at the beginning of the Canker War III, or in year 2000 when the 1900-ft policy was implemented, that anyone recognized how destructive the program would become in the end. However, there were some who suggested any eradication would be short-lived.[17] Was the citrus industry in better shape after the eradication program ended? This is hard to say. Today, it is likely that grove owners are much more concerned with the threat of citrus greening disease, which unlike canker, can cause a citrus tree to die.

It is still an open question of why did officials at both the USDA and FDACS continued to doggedly believe that there was a point at which the headline news would read, "Canker is forever gone from Florida." Throughout the eradication program, opponents pressed for changes in the program away from eradication and a new focus on living with canker. As Dr. Whiteside had explained during Canker War II, efforts to eliminate citrus canker are generally costly, laborious and the prospects for success are extremely dubious.[18]

FDACS was sued in a series of class action lawsuits for inadequate compensation of the healthy citrus trees. Robert Gilbert was the lead attorney in these cases. In Broward, Palm Beach, Orange and Lee counties, from 2011 to 2015, juries found in favor of homeowners, with awards totaling 57 million dollars. The victory in Palm Beach County was short lived as the court of appeals overturned the ruling of the trial court, citing the judge had incorrectly excluded evidence in the trial. The Department refused to pay in any case based on the law stating only the legislature can appropriate funds. Thus, the residents seem to lose their Fifth Amendment rights to just compensation for destruction of property, when this destruction comes at the hands of the State.

Beginning in 2006, details of the Citrus Health Response Plan (CHRP) and new regulatory rules began to emerge. The Plan was comprehensive, designed to manage citrus canker and other citrus diseases.[14] CHRP proposed a set of new regulations aimed at preventing widespread dissemination of these citrus diseases from nurseries to groves and residential outlets. To further assist grove owners, the UF/IFAS provides "best practices" recommendations to minimize the threat of canker and other diseases and pests.

2. USDA Restrictions on Exports of Fruit

The CCEP did not immediately end on January 10, 2006. The complete end to eradication required a change in Florida law. There were 2,329 square miles (1.5 million acres) under quarantine as of January 14, 2006. No citrus fruit or plants could legally be moved from a quarantine zone. In 2006, FDACS Commissioner Bronson lifted all restrictions on the sale of citrus trees within Florida. Because citrus canker had not been eradicated, there was no need for the two year waiting period before replanting.

There is a statewide quarantine in effect. This gives USDA authority to set strict requirements for inspection, sanitation and processing of citrus fruit. Inspectors call out any fruit with blemishes caused by mechanical damage, and bacterial and fungal diseases.

However, in June , 2006, the USDA proposed prohibition of export of Florida's citrus to other citrus producing states, including Texas, Louisiana, Arizona and California. This restriction resulted in strong criticism from Commissioner Bronson and Governor Bush.

Today, the entire state of Florida is living with canker. Fresh citrus fruit may be exported to all citrus producing states. The other countries such as Argentina, Mexico and Brazil do not need to worry that Florida might prohibit import of their citrus fruit, because are citrus fruit is grown in areas where citrus canker has been discovered. A series of USDA/APHIS studies determined that the packinghouse inspection and sanitary precautions were adequate to eliminate the risk that infected fruit could spread canker.[16-18] Dr. Schubert, of FDACS commented, "Even if infected fruit were to enter a canker-free area with susceptible hosts, the establishment of citrus canker via this pathway is highly unlikely." Thus, this "highly contagious disease" has been effectively downgraded as it is no longer needed for publicity purposes to support the CCEP.

This seems to be a complete reversal of the policy in place during the CCEP, and prior eradication efforts. During the CCEP, residents were warned that fruit from any citrus tree, within the 1900-ft eradication zone, had to be destroyed because it could spread the disease. When there was an economic need to export fruit, then studies were done to show citrus canker was much less contagious. Admittedly, the homeowners' measures to clean fresh fruit would not be the same as the regulated procedures in the packinghouses. Still there seem little chance residents' fresh citrus could cause new infections.

Opponents of the program were likely convinced that the Department knew better when it claimed that citrus canker was highly contagious. Some of the technical articles, referred to Asian citrus canker as the most virulent strain of citrus canker, without much evidence of how this virulence was determined.

Under current regulations, citrus trees grown in Florida can not be shipped to states with commercial citrus production (Arizona, Louisiana, Texas and California). This seems very reasonable requirement, as no matter how carefully the seedlings are examined, there is always a chance that a seedling with citrus greening, citrus canker or any other disease could be present in a shipment.

3. Central to South Florida Movement Theory

In Chapter 9, it was suggested that the canker disease originated primarily in nurseries in Central Florida. Hurricane Andrew in August 1992 created a demand for citrus, as four square miles of Key lime groves had been devastated. In the following years, contaminated seedlings were distributed to both coasts of South Florida. This would include both the pre-2000 discoveries in the area of Miami-Dade, Broward and Palm Beach Counties on the east coast of Florida and Collier, Hendry, Hillsborough and Manatee in the west coast. How the supply of citrus became contaminated is unknown. Possibly, the real culprits in the dispersal of canker were the delivery trucks making their way from Polk and Highlands counties, carrying nursery stock with hidden citrus canker infections to all areas within Florida. Some of these infected trees went to northern Florida, but environmental conditions were not ideal for canker. The epidemic was likely both a combination of re-introductions and remnants of the prior epidemic.

It is suggested that the Citrus Health Response Plan with its focus on the sanitation in the nurseries demonstrates the Department also was concerned. As noted in Chapter 9, the Comprehensive Report notes citrus canker was discovered in a large citrus nursery in Polk County in 2005. Seedlings were distributed to several groves. Contaminated seedlings could also been distributed to commercial outlets, and made their way into homeowners' backyards.

4. What Never Happened

For ten years, FDACS officials made dire claims of devastation of commercial grove production should the CCEP fail. Living with canker did not result in collapse of the citrus industry in Florida, nor an extensive ban on the export of citrus fruit outside of Florida. Exports to other countries continue. In the USDA 2013 to 2014 production report, Florida exported 3.4 million boxes of citrus to 19 countries. Of this total, 78% of the exports were shipped to Japan, Canada, Belgium and Holland. Exports account for 2.7% of Florida's production.

This acceptance of citrus was due in part to a study prepared by the USDA/APHIS, indicating that sanitation protocols within packinghouses were sufficient to reduce the chance of dissemination of canker to nearly zero. Also, the Citrus Health Response Program created a set of regulations for strict inspection of citrus in the packinghouses. The federal agencies (USDA/APHIS, USDA/ARS), state agencies (FDACS/DPI) and University of Florida/IFAS all partnered in this effort.

The four square miles of lime groves in the Homestead, Miami-Dade County were never replanted, despite compensation of millions of dollars given to grove owners for this purpose. Between year 2000 and 2002, a total of 76 million dollars were paid to grove owners. It was approximately half of the cost of the program at the time. They were free to sell their land to developers, and likely, given the real estate market in 2002 to 2006, many of them did sell out. The Office of Inspector General, provided further cases of grove owners who received substantial compensation and never replanted their lost citrus.

Full compensation for residents never occurred, nor was there a realistic manner to obtain their Constitutional rights. The Florida Supreme Court opined disgruntled residents could file against the Department in court. Class action lawsuits were filed and residents won in Broward, Palm Beach and Lee counties. However, as argued vehemently by Broward attorney Andrew Meyers during oral arguments in the Florida Supreme Court, the Department fought "tooth and nail" to deny residents any additional compensation.

The provisions for agricultural warrants still remain a part of Florida law. The law was revised to exclude the provision to permit application of a warrant which would cover the entire county. This has been ruled unconstitutional by the courts.

Finally, since the writings of Dr. Whiteside in the 1980's, it has been well established that buildings and trees are effective windbreaks. A busy highway, parking lots, canals and lakes would also impede dissemination of canker by windblown rain. Yet, no experimental studies have been conducted to show the effect of buildings to block the transport distances of windblown rain.

5. Citrus Health Response Program

The Citrus Health Response Program (CHRP) is focused on reducing the risk of introduction and dissemination of harmful citrus pests and diseases in Florida.[14] The USDA/APHIS and the FDACS/DPI worked jointly in the development of the program. The implementation of more extensive regulations and inspections for commercial nurseries and certified budwood facilities would likely be the responsibility of FDACS/DPI while interstate shipments would be regulated by the USDA/APHIS. Research aspects of the program would likely be the responsibility of the USDA/ARS, FDACS/DPI and University of Florida/IFAS through research grants.

As stated by the USDA/APHIS, the widespread citrus canker and citrus greening epidemics strongly motivated the development of the CHRP in 2006. However, the program is directed at all harmful pests and diseases which attack citrus plants. A total of 26 pests and diseases are listed in the 2006/2007 USDA Citrus Health Response Plan. Four diseases of major concern, as listed on the FDACS/DPI website, are: citrus greening, citrus black spot, citrus canker and citrus tristeza.

Nursery Regulations

One of the program's goal is to keep the nursery citrus as healthy as possible through increased regulations and inspections. Structures for both nursery stock and budwood sources must be grown in insect-resistant structures. There must be double entryway with positive pressure air displacement, so insects can not enter the structure.

The commercial citrus industry depends on "resetting" or the replacement of older citrus trees with seedling to maintain peak yields. As part of the CHRP management guidelines, the resetting can only be done with certified nursery stock and management strategies as defined in the compliance agreement. Other aspects address the production practices and harvesting within the commercial groves and post-harvest practices (packing of fresh fruit and processing facilities).

Newly established citrus nurseries must be located on an approved site that is a one mile distance from commercial groves. This is in accordance with Rule Chapter 5B-62 of the Florida Administrative Code. Existing citrus nurseries have no location restrictions in relation to commercial groves.

For a certified nursery, the surrounding area will be inspected for ½ mile around the site. This likely means entering the backyards of residents. If residents do not give permission, it is likely search warrants are required. If a citrus plant is found with citrus canker, FDACS/DPI has the authority to remove it, and all the fruit. It has no authority to remove healthy trees without full compensation.

Strict sanitation rules including decontamination of clothing for workers entering and exiting a nursery. Windbreaks for nurseries are not mandatory, but strongly recommended by FDACS/DPI. Complete details of the CHRP can be found on the USDA/APHIS and FDACS/DPI sites. Links to these sites are provided in the supporting documents website.

Citrus Greening (Huanglongbing) disease

Since the Citrus Health Response Program (CHRP) is directed at deterring or eliminating the most harmful pests and diseases to commercial citrus in Florida, it is necessary to include a brief mention of citrus greening disease. Citrus greening or Huanglongbing (HLB) disease was discovered in Florida in 2005. Citrus greening is a systemic bacterial disease which results in deform citrus fruit and ultimately can kill a citrus tree. The disease is carried and transmitted by the Asian citrus psyllid, a flying insect approximately 1/8 inch long. The Asian citrus psyllid was discovered in Florida in June 1998.

According to the USDA/ARS, "When a citrus tree is infected with the HLB bacterium, the pathogen resides inside the tree's phloem tissues and blocks the passage of nutrients through its vascular system, making the tree unproductive. Infected trees can survive for 3 to 5 years, but fruit that doesn't fall to the ground prematurely is often misshapen and sometimes will only partially ripen, making it unmarketable.

By 2007, citrus greening had spread throughout Florida. Since there is no cure, once it is positively identified, the recommended action is the removal of the infected tree. Initial symptoms of the disease resemble symptoms of common to nutritional deficiencies, so early detection is difficult. There is no known cure, so a tree with citrus greening must be destroyed, once it is positively identified. The most promising treatment appears to be to thermally treat citrus trees to temperatures above 100 degrees F. This must be done very carefully to avoid damaging the citrus. Research is continuing in this area as well as development of greening resistant cultivars.

Many excellent articles on citrus greening are publicly available. The UF/IFAS Citrus Research Education Center website as cited at the end of this chapter provides extensive information on citrus greening in Florida.

6. Citrus Canker Management Practices

The UF/IFAS provides a disease management guide for measures to reduce the risk of citrus canker introduction and dissemination within commercial groves. A link to their most recent guide is provided in the online supporting documents website. Many other publications on best practices for citrus groves are available online at the UF/IFAS website and are updated periodically to reflect advances in technology and horticultural practices. These recommended management practices have become much more critical to grove owners, now that the 1900-ft rule has ended, and compensation for their destroyed trees.

The guide suggests that in areas where citrus canker is endemic, recommend measures for control are (1) Windbreaks, (2) Protection of fruit and leaves with copper sprays, and (3) Control of citrus leafminer through insecticides. The most recent guide states the most effective means of prevention of citrus canker dissemination is through the planting of windbreaks. Research from UF/IFAS suggests that many other crops would also benefit from windbreaks. Suggested natural windbreak trees are red cedar, slash pine and eucalyptus.

Copper sprays have been suggested as a means of preventing citrus canker for more than 30 years. The articles published by Dr. Whiteside suggested the copper sprays could be more effective if properly timed.[21] The UF/IFAS also published online a guide for timing applications (EDIS PP-289, "A Web-Based Tool for Timing Copper Applications in Florida Citrus).[3] The copper spray program should also be applicable to homeowners, since these fungicides are environmentally safe and sold at commercial outlets.

For citrus leafminer (CLM), the management guide recommends the use of insecticides on the first summer flush for grove owners. Researchers at UF/IFAS have developed a "day-degree" computer model for timing the application of chemical sprays or synthetic pheromone based traps.[11] The online supporting website also contains many links to articles on citrus leafminer control. Traps are now available to residents at nursery outlets. Neonicotinoid insecticides have become increasingly popular, but are have become very controversial as to their affect on bees discussed in the next section.

In protecting canker free areas, the management guide recommended measures are: (1) Decontamination of equipment using in operations, including harvesting and hedging/topping of trees , (2) Removal and burning of any infected tree, with possible removal of healthy trees immediately surrounding area and (3) Defoliation and pruning of infected trees.

The above recommendations for removal and defoliation would be options for grove owners (and presumably also for home owners) when fruit or foliage is confirmed infected with citrus canker. While healthy tree removals are suggested, there is no exposure radius suggested, such as 125-ft, or 1900-ft. In fact, an internet search of articles from the USDA/APHIS, FDACS/DPI and UF/IFAS, showed no publication recommending eradication of healthy trees within a predetermined distance from an infected one.

So, each grove owner will have to evaluate whether to remove or defoliate infected trees, and whether to remove healthy trees as well. If the grove owner decides to remove healthy trees, the next question is how far within the row and between rows should be removed.

There are no reports of what actions grove owners are taking today when citrus canker is discovered. During the CCEP, the FDACS frequently stated that the grove owners were in full support of the 1900-ft eradication circles. However now, when the compensation program no longer exists, and removal of healthy trees not obligatory, what are the grove owners actually doing?

Certainly, a discovery of citrus canker in a seedling early in the summer months might suggest removal rather than defoliation. Conversely, canker in an older tree in the winter months might suggest defoliation rather than removal. In either cases, copper sprays and monthly inspections would be the recommended actions for grove owners with canker discoveries.

7. Use of Neonic Insecticides (Citrus Health = Bee Death?)

Systemic neonicotinoid insecticides (commonly referred to as neonics) are used against many pests including the citrus leafminer and the Asian citrus psyllid. They are widely available in retail stores. The popularity of neonicotinoid compounds is due to the fact that these chemicals provide a long lasting, economical option to eliminate many pests including adelgids, aphids, black vine weevil larvae, borers, beetles, lacebugs, leafhoppers, leafminers, mealybugs, pine tip moth larvae, psyllids, royal palm bugs, sawfly larvae, scale, thrips and whiteflies.[24] For pests in Florida, it is commonly used as a drench.

The best selling neonicotinoid pesticide is imidacloprid with over one billion dollars in sales.[25] It is manufactured by Bayer AG. According to Wikipedia,

> Imidacloprid is currently the most widely used insecticide in the world. Although it is now off patent, the primary manufacturer of this chemical is Bayer CropScience (part of Bayer AG). It is sold under many names for many uses; it can be applied by soil injection, tree injection, application to the skin of the plant, broadcast foliar, ground application as a granular or liquid formulation, or as a pesticide-coated seed treatment. Imidacloprid is widely used for pest control in agriculture. Other uses include application to foundations to prevent termite damage, pest control for gardens and turf, treatment of domestic pets to control fleas, protection of trees from boring insects, and in preservative treatment of some types of lumber products (e.g., Ecolife brand). It is huge and complex controversy because these insecticides have been considered safe alternatives to other chemicals, and are very widely used in farms throughout the country against a very wide varieties of harmful insects. " [25]

Other neonicotinoid compounds are commercially available including clothianidin and thiamethoxin.

Neonicotinoids are neurotoxins which can be ingest by bees through the pollen and nectar. The concentrations of the ingested pesticide are typically at sub lethal levels to bees. However, recent studies suggest the bees ability to forage for nectar may be impaired through the ingestion of extremely low concentrations of neonicotinoids.[10,24] The threat is to the colony of bees, rather than individual bees. In a laboratory setting, the impairment may not be observed or be subtle, but in a field settings, experiments have shown the intake of neonicotinoids can produce profound consequences to the survival of the entire colony. In 2013, the European Union and a few non EU countries restricted the use of certain neonicotinoids. The environmental safety of certain neonicotinoids will be evaluated by the US Department of Environmental Protection Agency in 2016- 2017 as part of the renewal of registration process. [25]

The environmental effects of neonicotinoids are presently undergoing intense investigations worldwide. Excellent summaries of the history and present studies can be found on Wikipedia. In June 2014, the White House authorized a Pollinator Health Task Force, to be co-chaired by the USDA and the EPA. [23] Their report, entitled National Strategy to Promote the Health of Honey Bees and Other Pollinators, chronicled the decline in both Monarch butterflies and honey bees. The report suggests that pesticide use could have been, in part, responsible for this decline. [23] Advocates of a ban on neonicotinoids may have been disappointed as the Task Force did not directly identify neonicotinoid usage as responsible for the decline in pollinators. The Task Force report shows "pollinator specific budget additions for 2016" will be 34 million dollars, of which the EPA's Office of Pesticides will receive 1.5 million dollars. Most of the additional budgeted funds will go to the USDA.

The EPA is the final arbiter for the registration of pesticides. On January 4, 2016, the EPA, Office of Chemical Safety and Pollution Prevention, issued a report entitled, "Preliminary Pollinator Assessment to Support the Registration Review of Imidacloprid" to provide a solid technical review of the affects of a specific pesticide (imidacloprid) on a particular insect (bees). [20] The primary focus of the risk assessment was for honey bees (*Apis mellifera*), but where the data were available, affects on non-*Apis* bees were considered. The approach is explained on page 24 of the report as follows, "The decision to focus on imidacloprid's potential risk to bees (honey bees [*Apis mellifera*] and non-Apis bees) reflects that Agency's desire to evaluate potential risks and appropriate mitigation measures earlier in the Registration Review process relative to other taxa. It also reflects the large volume of information related to environmental exposure and effects of imidacloprid to bees which has been generated over the past decade."

The targeted strategy on the impact of imidacloprid on bee colony survival, addresses the primary concerns of environmental groups. However, the risk assessment is still multi-faceted, as imidacloprid is used to treat so many different crops, and may be applied as a foliar spray, drench of seed coating. Other environmental effects on colony survival need to be accounted for, such as mites or other common predators of bees.

The preliminary assessment presented in the report[20] shows significant risk to honey bees, when imidacloprid is used to treat citrus fruits (see Table 1-2 of report). The risk is higher when imidacloprid is used as a foliar spray. Reports are open to public comment and it is likely Bayer A.G. and agricultural representatives will respond. The White House Task Force reports and the EPA report are posted on the supporting documents website.

8. Florida's Citrus - Past and Present

... [in the future, after eradication] residents can replant citrus trees that will bear healthy, fresh fruit for themselves, their children and grandchildren to enjoy.

Commissioner of FDACS Bob Crawford. Brochure distributed by FDACS prior to implementing the 1900-ft rule in year 2000.

"From now on, we're going to have to learn to live with it", Charles Bronson, October 6, 2006.

The promise of a canker-free Florida, of course, was never realized. By 2006, 88 thousand acres of commercial groves were gone, and over a billion dollars spent.

A far worse disease, citrus greening, was discovered in 2005, and quickly was disseminated throughout Florida.[7] Government agencies are doing everything possible to find a solution. This includes efforts in the US, Brazil and all other citrus producing countries.

Fresh fruit will still be available for our children and grandchildren in Florida, with or without citrus canker. Perhaps not as abundant or inexpensive, but citrus is not disappearing from Florida. It will be here, together with citrus greening, citrus black spot, oriental fruit flies, and a host of other pests and diseases.

The following sections provide brief summaries of citrus production and related statistics for the last 15 years as obtained from USDA-National Agricultural Statistics Service and the USDA Florida Citrus Statistics Report. Short Note 10.1 provides further details and references on Florida's citrus.

Citrus Production

Both production and acreage have declined from 1998 to 2015. It is a continuing trend that has existed for three decades. As of 2015, Florida still has approximately half a million acres of citrus groves. The USDA, National Agricultural Services (NASS), shows a nearly 30 year downward trend in the citrus bearing acreage and boxes of citrus fruit produced in Florida. The latest report shows citrus bearing acreage to be 427 thousand acres and the ninth year of decline. The preliminary estimate of citrus production in Florida in 2015 from the USDA is 113 million boxes, the lowest since the 1963 to 1964 growing season and roughly half of the production in 1992- 2004 period.

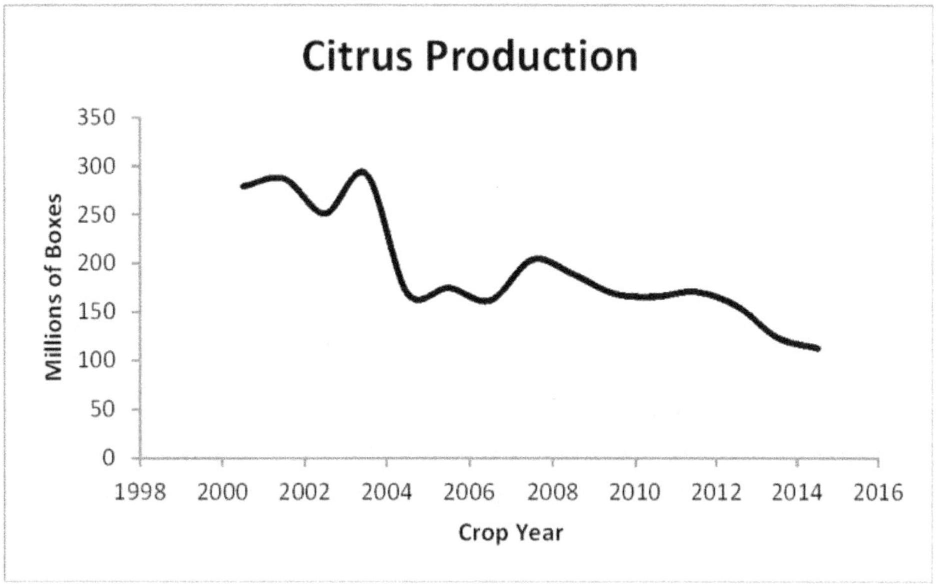

.

Figure 10.1 Citrus Production

On-Tree Value/Box

The USDA Florida Citrus Statistics Report provides an approximate measure of the direct benefits the grove owner receives for the sale of citrus. The term "on-tree" relates to fruit returns to the grower after the costs of picking, hauling, and packing have been removed. The on-tree value is not an estimate of net cash flow, as there are also other expenses such as administrative overhead excluded from this value.

The on-tree value/ box of citrus in year 2000 was $3.26/box, while in year 2015, it is $9.18/box. A box of oranges weighs 90 pounds, while a box of grapefruit weighs 85 pounds. The upward trend in on-tree value per box is shown in Figure 10.2.

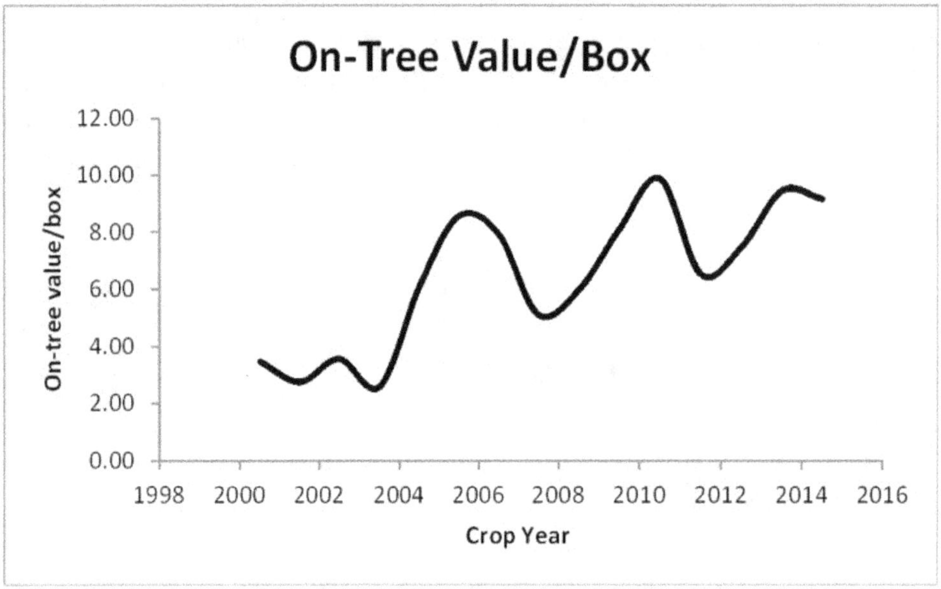

Figure 10.2 On-Tree Value per Box

Both citrus acreage and production shows a similar declining trend in the last decade. However revenues as measured based on the on-tree value per box of citrus, have an upward trend. So, on balance, fruit price increases have compensated for production declines over the last 15 years.

The total on-tree value of Florida's citrus groves was 966 million dollars in 2000 to 2001 while the value in 2014 to 2015 growing season is estimated to be 1034 million. This is not to suggest that the on-tree value trend line will be positive in the future. The on-tree value from the peak value of 1647 million in crop year 2010-2011 and then declined by 37% to 1034 million in crop year 2014-2015.[9] The upward trend in on-tree value and the declining trend in production is shown in Figure 10.3.

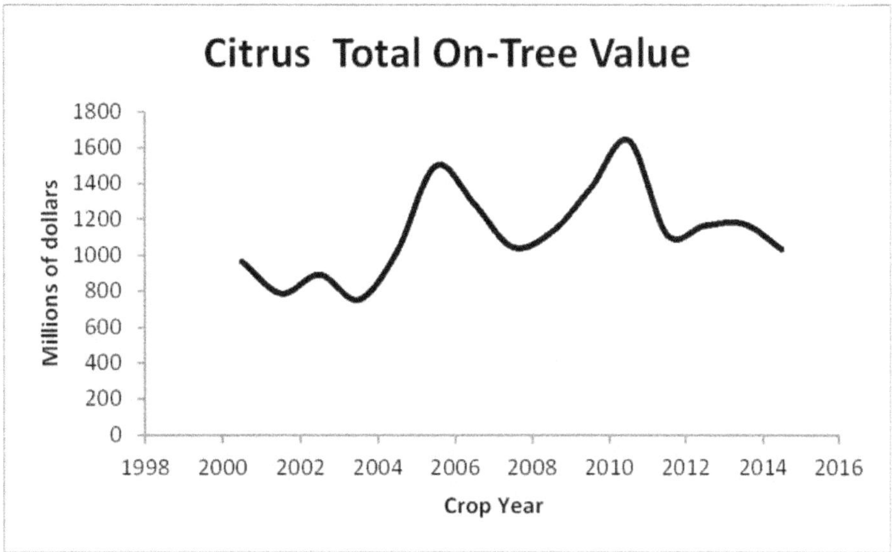

Figure 10.3 Citrus Total On-Tree Value

The increased value of citrus has offset the diminished production since year 2000, with an increased on-tree value of citrus groves of 6.6% in the last 15 years.[19]

Newly Planted Acreage

Aerial surveys were conducted on even numbered years, until year 2008 when the surveys were done every year. The planted acres have been converted to an annual basis for comparison purposes. Peak newly planted acreage occurred in 1992 with a total of 66,614 acres. A rapid drop off occurred between 1994 to 1996, as the new planted acres dropped from 53,818 to 18,946 acres.

The CCEP destroyed 11.3 million commercial citrus trees for a total of 87,493 acres. The general decline in newly planted acreage after year 2000 and continuing beyond 2006, is an indication that the grove owners have either switched to other crops or sold their groves to developers, rather than replant citrus. It is estimated that approximately half of the 1.3 billion dollars spent on the eradication program, was for compensation to grove owners, so they would re-plant and keep the citrus industry alive. It certainly does not seem this was money well spent.

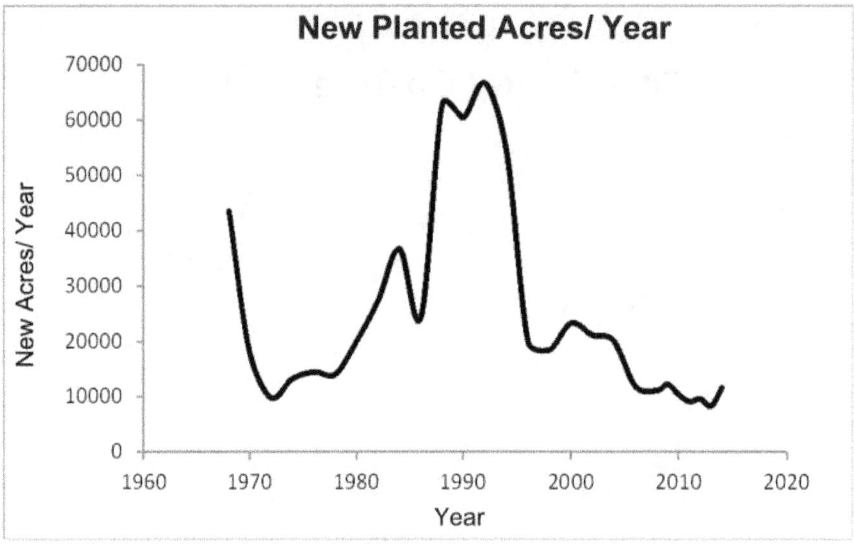

Figure 10.4 Newly Planted Acres per year

The Future

The future should be best left up to better prognosticators than myself. With time, the trends as shown on Figures 10.1 to 10.4 will fade into history. However, the future appears to be neither bleak nor bright. The disappearance of groves is no mystery, at least for anyone who drives through agricultural areas with large "For Sale" signs from home builders. Farm land prices close to urban areas soared in the housing boom years (2000 to 2007) due to economics and the ease of conversion. In the past, grove owners have reduced tree spacing to make up for lost acreage. It is likely this can not continue in the future without sacrificing yield. Less land means fewer oranges and grapefruits grown in Florida and higher dependence on fresh fruit from other countries. The consumer is likely to be hurt more than the grove owner.

There will be a dynamically changing environment, both for consumers and producers. The popularity of Not From Concentrate (NFC) juice may improve prospects for grove owners. One recent book on the subject, has been critical of this innovation, concerned that consumers may not be aware that this juice is also the result of a sequence of processing steps.[8] Residents may be more motivated to plant their own citrus, particularly given the high price of citrus fruit in the stores. A recent study suggests that commercial lime production may be now profitable in South Florida despite the problems with pests and diseases.[5]

9. Concluding Remarks

"The reports of the citrus industry demise have been greatly exaggerated." Pardon me for adapting Mark Twain's famous quote about a premature announcements of his own demise to the citrus industry. The citrus industry did not fall into ruin and disappear after the canker eradication program ended. Through the Citrus Health Response Plan, the Department has a sufficiently flexible plan to manage citrus canker and other diseases, such as greening.

The CCEP did more harm than waste more than a billion dollars and destroy 16.5 million citrus trees. It created a feeling of mistrust for the FDACS. The refusal of the Department to pay the judgments of the courts for just compensation is an affront to the rights of resident as guaranteed in Fifth Amendment of the Constitution. Adopting a policy to pay fair compensation for fruit and trees would improve trust in the Department.

New challenges lie ahead. The EPA will face a daunting decision when it makes a decision on the continued registration of neonicotinoids. The EPA will need to examine impact of neonicotinoids on honeybees at the colony level. If done properly, this will require extensive field experiments with many different crops. It is a complex and unenviable task.

Citrus will continue to survive in Florida. It is not citrus canker, citrus greening or fruit flies that pose a the long term threat for Florida's commercial citrus. "We have met the enemy and he is us," to quote Walt Kelly, author of the cartoon strip Pogo. Or more correctly, the enemy is our desire for new homes away from the urban sprawl, which will continue to reduce the commercial citrus acreage. Low unemployment and interest rates likely will drive the demand for housing up, and continue the shrinkage of available farm land in Florida. Fresh fruit will still be in Florida, just at higher prices.

In this environment of increasing commodity prices, it is likely homeowners will be planting their own fruit trees. Perhaps, more orange and grapefruits will come from residents' backyards. Maybe more vegetable and herb gardens will accompany their fruit trees. Economically it makes sense.

Short Notes on the Website:

SN 10.1 Commercial Citrus Production and Related Measures

Notes

Chapter 1: Citrus Canker Eradication Program

1. Anonymous, "Bahamas fighting citrus canker outbreak", Caribbean News Network, January 24, 2005.
2. Canteros, B.I., 2000. Citrus Canker in Argentina — Control, Eradication and Current Management, International Citrus Canker Workshop, June 20-22, 2000. (Abstract and transcript only),
3. Canteros, B.I., 2004, Citrus Canker in Argentina— A Review, Proc. Int. Soc. Citriculture, Paper No. 9, 2004.
4. FDACS/DPI, 1999, Citrus Tree Value, internal memo dated March 1, 1999 from Connie Riherd to Richard Gaskalla. (copy available online).
5. FDACS/DPI, 1999, Citrus Canker in Dade and Broward Cos. Report No. CCRAG-9, May 11, 1999, Minutes of Meeting. Submitted into evidence in Broward 17th District Court, in November 2000 (Case 00-18394 (08).
6. FDACS/DPI, 2003, Plant Plant Pathology Circular 377, Bacterial Citrus Canker, Schubert, T.S. and Sun, X.
7. FDACS, 2006. USDA Declares Citrus Canker Eradication Infeasible, Press Release from Commissioner Charles Bronson, January 11, 2006.
8. FDACS, 2012. Comprehensive Report in Citrus Canker in Florida, Florida Department of Agriculture and Consumer Services, 2012.
9. Gottwald, T. R. and Graham, J. H., Schubert, T.S. ,1997. An Epidemiology Analysis of the Spread of Citrus Canker in Urban Miami, Florida and Synergistic Interaction with the Asian Citrus Leaf Miner, Fruits, Vol 52-5, 371, 1997.
10. Gottwald, T.R. and Graham, J.M., 1999. Research in Support of Eradication and Control of Asiatic Citrus Canker, Project 981-29. Year 1 progress report.
11. Gottwald, T. R., Sun, X., Riley, T., Graham, J., 1999. Estimating the Spread of Citrus Canker in Urban Miami via GPS, Annual APS Meeting Abstract, Phytopathology, 89:S29. Publication no. P-1999-0202-AMA (Poster Presentation 1999 meeting, Montreal).
12. Gottwald, T. R. and Graham, J.H., 1999. Research in Support of Eradication and Control of Asiatic Citrus Canker (ACC), July 1998- June 1999 Annual Report. (online copy available, http://citrusrdf.org/annrep/1999-ann-rep/fcpracarep99.htm)
13. Gottwald, T. R., Sun, X., Riley, T., Graham, J., 2000. Interaction of meteorological events with the increase and spread of citrus canker, Annual APS Meeting Abstract, Phytopathology, 90:S29. Publication no. P-1999-0207-AMA (2000 meeting, Montreal).
14. Gottwald, T. R., Hughes, G., Graham, J. H, Sun, X., Riley, T., 2001, The Scientific Basis of Regulatory Eradication Policy for an Invasive Species, Phytopathology, 91:30-34.
15. Gottwald, T.R., Graham, J.H., Schubert, T.S., 2002, Citrus Canker: The Pathogen and Its Impact, Plant Health Progress, published online at www.apsnet.org (official website of the American Phytopathological Society).

16. Gottwald, T.R., X. Sun, Riley, T. Graham, J. H., Ferrandino, F. and Taylor, E., 2002, Geo-Referenced Spatiotemporal Analysis of the Urban Citrus Canker Epidemic in Florida, Phytopathology, Vol 92, No. 4. (An average of the 5 sites showed approximately 2000 citrus/sq mile).
17. Gottwald, T.R., 2005, Citrus Canker, The Plant Health Instructor, Online Article, American Phytopathological Society, initially posted year 2000, updated in 2005.
18. Gottwald, T.R., 2007, Citrus Canker and Citrus Huanglongbing, Two Exotic Bacterial Diseases Threatening the Citrus Industry of the Western Hemisphere, Outlooks on Pest Management, October 2007. Online article, see link in online supporting documents website.
19. Graham, J.H., 1998, "Citrus Canker, Control Efforts in Brazil, Prognosis for Florida", 1998, Citrus Industry, **79**(8), 54-57.
20. Graham, J.H., Gottwald, T.R., Cubero, J. and Achor, D. S., 2004, Xanthomonas axonopodis pv. citri: factors affecting successful eradication of citrus canker, Molecular Plant Pathology, 5(1), 1-15.
21. Jetter, K.M, Summer, D.A. and E.L. Civerolo, "Ex Ante Economic Disease Policy on Citrus Canker in California", presented at the Integrating Risk Assessment Economics for Regulatory Decisions, USDA, Washington.
22. LaVigne, A., 2000, Opening Remarks on June 20, 2000, International Citrus Canker Research Workshop, page 15-16 of transcript.
23. Lowe, D., 2009, Current Situation, Management, and Economic Impact of Citrus Canker in Florida, USDA/APHIS/PPQ, Sept 2009 (online copy available).
24. Nielsen, K. , 2000, Anatomy of a Quarantine, Miami New Times, July 6, 2000.
25. Riley, T., Current Situation, Management, and Economic Impact of Citrus Canker in Florida, USDA/APHIS/PPQ, Sept 2009 (slide presentation, online copy available).
26. Schubert, T.S. , Gottwald, T.R., Rizvi, S.A., Graham, J.H., Sun, X., Dixon, W.N., 2001. Meeting the Challenge of Eradicating Citrus Canker in Florida- Again, Plant Disease, Vol. 85-4.
27. Taylor, Jean C. The Citrus Canker, Historical Museum of South Florida (online copy available).
28. USDA/APHIS, 1999, Citrus Canker Eradication Program: Environmental Assessment, April 1999. Agency contact listed as Dr. Stephen Poe, Program Officer.
29. USDA, Office of Inspector General, South East Region, 2002. Audit Report: APHIS Citrus Canker Eradication Program State of Florida, Report 33099-2-At, August 2002.
30. USDA memorandum, 2006 USDA Deputy Secretary, Chuck Conner to FDACS Commissioner Charles Bronson, January 10, 2006.
31. USDA, APHIS- Office of Inspector General, 2011. USDA Payments for Citrus Canker Payments, March 23, 2011.
32. Whiteside, J.O., "How Serious a threat is canker to Florida's Citrus Production, Citrus and Vegetable Journal, 1985.
33. Whiteside. J.O., September 1986, Citrus Canker, Some Facts, Speculations and Myths about this Highly Dramatized Bacterial Disease, Citrus and Vegetable Magazine.

34. Whiteside, J.O., April 1988, "The History and Rediscovery of Citrus Canker in Florida, Citrus and Vegetable Journal.
35. Zansler. M.L., Spreen, T.H. and Muraro, R.P.,2005. Florida's Citrus Canker Eradication Program (CCEP): Benefit-Cost Analysis, FE531, University of Florida, IFAS Extension, February 2005. Also see FE 532, Summary of Annual Costs and Benefits by Zansler et al.

Supplemental Notes

36. Citrus Canker Risk Assessment Group, 1999. Bacterial Citrus Canker and the Commercial Movement of Fresh Citrus Fruit, July 14, 1999. (only a draft copy is available).
37. Citrus Canker Pathology Training Presentation, Citrus Health Response Program, Texas A+M, (provided on website).

Chapter 2: Four Decades of Eradication

1. FDACS/DPI, 2003, Plant Plant Pathology Circular 377, Bacterial Citrus Canker, Schubert, T.S. and Sun, X.
2. Graham, J.H., 1998, "Citrus Canker, Control Efforts in Brazil, Prognosis for Florida", 1998, Citrus Industry, 79(8), 54-57.
3. Jalan, N., Aritua, V., Kumar, D., Yu, F., Jones, J., Graham, J.H., Setubal, J. Wang, N., 2011. Comparative Geonomic Analysis *Xanthomonas axonopodis* pv. citrumelo F1 Which Causes Citrus Bacterial Spot Disease and Related Strains Provides Insights into Virulence and Host Specificity, Journal of Bacteriology, 2011, Vol 193, No. 22.
4. Lakeland Ledger, November 1, 1994. Nurseries may have exported citrus canker
5. Lakeland Ledger, Aug 25, 1985, Ward confident of canker's defeat.
6. Longman, P., 1989. The Big Lie, Florida Trend Magazine, February 1989.
7. Nielsen, K. , 2000. Anatomy of a Quarantine, Miami New Times, July 6, 2000.
8. Observer Reporter, September 24, 1984, More Burning will Follow, Associated Press.
9. Palm Beach Post, Jan 27, 1916. Government will expend $300 in eradication of citrus canker
10. Palm Beach Post, May 25, 1919. Sentiment to Divide Florida Grows Strong When Measure is Introduced in Legislature by Mathis of Holmes. (An Amendment to divide Florida into two parts, Florida and South Florida was introduced in the Florida House of Representatives).
11. Sarasota Herald, January 11, 1988. New Data offered on Citrus Canker.
12. Schubert, T.S. , Gottwald, T.R., Rizvi, S.A., Graham, J.H., Sun, X., Dixon, W.N., 2001. Meeting the Challenge of Eradicating Citrus Canker in Florida- Again, Plant Disease, Vol. 85-4.
13. Stevens, H. E., 1915. Nature and cause of citrus canker, Proc. Fla. State. Hort. Soc., 1915, pp .81-85
14. USDA/APHIS, 1999, Citrus Canker Eradication Program: Environmental Assessment, April 1999. Agency contact listed as Dr. Stephen Poe, Program Officer.
15. Whiteside, J.O., 1985. How Serious a threat is canker to Florida's Citrus Production, Citrus and Vegetable Journal, 1985.

16. Whiteside. J.O., September 1986. Citrus Canker, Some Facts, Speculations and Myths about this Highly Dramatized Bacterial Disease, Citrus and Vegetable Magazine.
17. Whiteside, J.O., April 1988. "The History and Rediscovery of Citrus Canker in Florida, Citrus and Vegetable Journal.
18. Wikipedia link- http://en.wikipedia.org/wiki/1906_Florida_Keys_hurricane

Chapter 3: Biology

1. Belasque et al., Adult Citrus Leafminers (Phyllocnistis citrella) Are Not Efficient Vectors of Xanthomonas axonopodis pv citri, Plant Disease, Vol 89, June 2005.
2. Blanke, M.M., 1996, Stomata, Transpiration and Photosynthesis of Citrus Orange Fruit, Proc. Int. Soc. Citriculture, 1996, Vol. 2.
3. Canteros, B.I., Citrus Canker in Argentina- Control, Eradication and Current Management, International Citrus Canker Research Workshop, June 20-22, 2000. Similar comments in Proc. Int. Soc. Citriculture, 2004, Paper No. 9, Management of Citrus Canker in Argentina, A review.
4. Civerolo, E.L., 1981. Citrus Bacterial Canker Disease: An Overview, Proc. Int. Soc. Citriculture, 1981, Vol 1.
5. FDACS/DPI , 2001. Responses to questions asked at Public Hearing on November 14, 2001. http://www.freshfromflorida.com/content/download/10005/136298/public-hearing11-14-01.pdf. question #30.
6. FDACS, 2012, Comprehensive Report in Citrus Canker in Florida, Florida Department of Agriculture and Consumer Services, 2012.
7. Florida Fruit and Vegetable Association, 2000, Search and Destroy, Florida Steps Up War on Citrus Canker. May/June 2000. (Comments by DPI Director Richard Gaskalla)
8. Gottwald, T.R., Graham, J.H., Egel, D.S.,1992, Analysis of Foci of Asiatic Citrus Canker in a Florida Citrus Orchard, Plant Disease, April 1992.
9. Gottwald, T.R., Graham, J.H. and Schubert, T.S. 1997. An Epidemiology Analysis of the Spread of Citrus Canker in Urban Miami, Florida and Synergistic Interaction with the Asian Citrus Leaf Miner, Fruits, Vol 52-5, 371, 1997.
10. Gottwald, T.R. Graham, J.H., and Schubert, T.S., 1997. Citrus Canker in Urban Miami: An analysis of spread and prognosis for the future, Citrus Industry, 78m, 72078. 1997.
11. Gottwald, T.R., X. Sun, Riley, T. Graham, J.H., Ferrandino, F. and Taylor, E., 2002. Geo-Referenced Spatiotemporal Analysis of the Urban Citrus Canker Epidemic in Florida, Phytopathology, Vol 92, No. 4.
12. Gottwald, T.R. and Graham, J.H., 2002. Citrus Canker: The Pathogen and Its Impact, August/September 2002, Plant Health Progress. Online copy, at www.apsnet.org.
13. Graham, J., and Gottwald, T.R., 2000. Survival of Xanthomonas campestris pv. citri on Various Surfaces, International Citrus Canker Research Workshop, June 20, 2000. (Dr. Gottwald presented p. 70)

14. Graham, J.H., Gottwald, T.R., Cubero, J. and Achor, D. S., 2004, Xanthomonas axonopodis pv. citri: factors affecting successful eradication of citrus canker, Molecular Plant Pathology, 5(1), 1-15.
15. Heppner, J.B, 1995. Citrus Leafminer (Lepidoptera: Gracillariidae) on Fruit in Florida, Florida Entomologist, March 1995.
16. Jeble, R.A., 1916. Means of Identifying Citrus Canker, The Quarterly Bulletin, State Plant Board of Florida.
17. Jackson, L.K. and Davies, F.S., Citrus Growing in Florida, University Press of Florida, 1999.
18. Jetter, K.M, Sumner, D.A. and Civerolo, E. L., 2000. Ex Ante Economic Disease Policy on Citrus Canker in California, presented at the Integrating Risk Assessment Economics for Regulatory Decisions, USDA, Washington.
19. Kumar, N., Egel, R., Roberts, P.D., 2013. Canker resistance: lessons from kumquats, Citrus Industry, February 2013.
20. Miami Herald, November 5, 2000. Canker Hotline shows breadth of resident's ire 22,609 complaints logged in in 13 months.
21. Observer-Reporter, PA, Sept 20, 1984. More Burning will follow. Associated Press, Lakeland.
22. Presentations at the International Citrus Canker Research Workshop, June 20, 2000. Gottwald, T.R. Sun, X., Riley, T., Hughes, D., Graham, J. "Estimating Spread of Citrus Canker in Urban Miami via Differential GPS", p. 34-54, Schubert, T.S, Sun, X., Dixon, W., Integrating the Scientific Perspectives in Citrus Canker Eradications Efforts in Florida, p. 138- 158. Also see references 3, 13, and 34 for other presentations.
23. Pruvost, O., Bober, C., Brocherieux, C., Nicole, M., Chiroleu, F., 2002, Survival of Xanthomonas axonopodis pv citri in Leaf Lesions Under Tropical Environmental Conditions and Simulated Splash Dispersal of Inoculum, Phytopathology, Vol 92, April 2002.
24. Roistacher, C.N. and Civerolo, E.L., 1989, Citrus Bacterial Disease of Lime Trees in Maldives Islands, Plant Disease, April 1989.
25. Schubert, T.S., Miller, J.W., Dixon, T.R., Gottwald, T.R., Graham, J.H., Hebb, L.H., Poe, R.R., 1999. Bacterial Citrus Canker and the Commercial Movement of Fresh Citrus Fruit, draft report only, July 14, 1999. Prepared by the Citrus Canker Risk Assessment Group.
26. Schubert, T.S, Rizvi, S.A., Sun, X., Gottwald, T.R, Graham, J.H., Dixon, W.N., Meeting the Challenge of Eradicating Citrus Canker in Florida- Again. Plant Disease, April 2001.
27. Schubert, T.S. and Miller, J.S., Bacterial Citrus Canker, FDACS Plant Pathology Circular 377. Fifth Revision. 2003.
28. Seriwaza, S. , 1981, Recent studies on the behavior of the casual bacterium, Proceedings of the International Society of Citriculture, 1981, 395-7.
29. Stall, R.E, Marco, G.M., and Canteros de Echenique, B.I., Population Dynamics of Xanthomonas citri causing cancrosis of citrus in Argentina, Proc. Fla. Hort. Soc., 93: 10-14.
30. Stapleton, J.J., 1991, Citrus Leaf Spot in Mexico- A New Alternaria Disease, South of the Border, Cooperative Extension, University of California, Kearney Plant Protection Group, Plant Protection Quarterly, April 1991.

31. Sun, X, Stall, R. Jones, J., Cubero, J., Gottwald, T., Graham, J., Dixon, W.N., Schubert, Peacock, M., Sutton, B., Dickstien, E., Chaloux,P., Detection of a unique isolate of Citrus Canker Bacterium in Wellington and Fort Worth, Florida. International Citrus Canker Research Workshop, June 20, 2000. (only abstract and transcript of talk available)
32. Sun, X., Stall, R.K, Jones, J.K., Cubero, J., Gottwald, T.R., Graham, J.H., Dixon, W.N., Schubert, T.S., Stromberg, V.K., Lacy, G.H, Sutton, B.D., Detection and Characterization of a New Strain of Citrus Canker Bacteria from Key/Mexican Lime and Alemow in South Florida, Plant Disease, November 2004.
33. Swings, J.G and Civerolo, E.L., Xanthomonas, 1993, Springer. ISBN 978-94-011-1526-1
34. Timmer, L., 1996. Citrus Leafminer Proves to be an IPM Success, reprinted on the web by permission of Florida Growers and Ranchers, Nov 1996.
35. Timmer, L., 2000. "Inoculum Production of Epiphytic Population Survival of Xanthanomos campestris pathovar citrus", International Citrus Canker Workshop, June 20-22, 2000. (Posted to FDACS website, no paper only abstract and transcript available)
36. Timmer, L.W., Zitko, Gottwald, T.R., S.E., Population Dynamics of Xanthomonas campestris pv. citri on Symptomatic and Asymptomatic Citrus Leaves under Various Environmental Conditions, Proc. Int. Soc. Citriculture, 448- 451, 1996.
37. University of Florida - IFAS, 2014. Florida Citrus Pest Management Guide: Citrus Canker, Dewdney, M.M. and Graham, J.H., available online.
38. USDA/APHIS, June 2007, Movement of Commercially Packed Citrus Fruit from Citrus Canker Disease Quarantine Area. Available online.

Supplemental Notes:

39. Center of Agricultural and Bioscience International website, (CABI) Datasheet from the Invasive Species Compedium, www.cabi.org/isc/datasheet/5691
40. Correspondence, Mr. Richard Gaskalla, Review of Mr. David Lord's review of the epidemiology study, December 2001.
41. Tucker, D.P.H., Wheaton, T.A. and Muraro, R.P., Citrus Tree Spacing, University of Florida, Florida Cooperative Extension Service, Fact Sheet HS-143.
42. Citrus Canker Risk Assessment Group, 1999. Bacterial Citrus Canker and the Commercial Movement of Fresh Citrus Fruit, July 14, 1999. (only a draft copy is available).
43. Citrus Canker Pathology Training Presentation, Citrus Health Response Program, Texas A+M, (provided on website).
44. Behlau, F., Barelli, N.L., and Besasque, J., 2014. Lessons from a Case of Successful Eradication of Citrus Canker in a Citrus-Producing Farm in São Paulo State, Brazil, Journal of Plant Pathology, (2014), 96 (3), 361-368.
45. Graham, J.H., Gottwald, T.R. and Achor, D., 1992. Penetration Through Leaf Stomata and Growth of Xanthomonas campestris in Citrus Cultivars Varying in Susceptibility to Bacterial Diseases. Phytopathology 82:1319-1325.

46. Schubert, T. and Sun, X., 2000. Integrating the Scientific Perspective into Citrus Canker Eradication Efforts, International Citrus Canker Research Workshop, June 20, 2000, page 138.

Chapter 4: Opposition to the Eradication

1. Graham, J.H., 1998. "Citrus Canker, Control Efforts in Brazil, Prognosis for Florida", Citrus Industry, **79**(8), 54-57.
2. Gottwald, T.R., Graham, J.H. and Schubert, T.S., 1997. An Epidemiology Analysis of the Spread of Citrus Canker in Urban Miami, Florida and Synergistic Interaction with the Asian Citrus Leaf Miner, Fruits, Vol 52-5, 371, 1997.
3. Gottwald, T.R. and Graham, J.H., 2002, Citrus Canker: The Pathogen and Its Impact, August/September 2002, Plant Health Progress. Online copy, at www.apsnet.org.
4. FDACS Press Release, 1999. Citrus Canker Eradication Project to Resume Cutting Exposed Trees, June 17, 1999.
5. FDACS/DPI , 2001, Responses to questions asked at Public Hearing on November 14, 2001.
6. Miami Herald, 1999. Florida Citrus Canker Inspectors will Start Revisiting Yards, June 19, 1999.
7. Schubert, T.S. , Gottwald, T.R., Rizvi, S.A., Graham, J.H., Sun, X., Dixon, W.N., 2001, Meeting the Challenge of Eradicating Citrus Canker in Florida- Again, Plant Disease, Vol. 85-4.

Chapter 5: Legal Challenges

The opinions of the court as cited in this chapter have been provided on the website, www.citruscankerdocs.com along with comments by key participants in these cases.

Chapter 6: Selected Topics in Epidemiology

1. Bock, C.H., Parker, P.E.,Gottwald,T.R., 2005. Effect of Simulated Wind-Driven Rain on Duration and Distance of Dispersal of *Xanthomonas axonopodis* pv. *citri* from Canker-Infected Citrus Trees, Plant Disease, Vol 89. No. 1.
2. Bock, C.H., Graham, J.H., Gottwald,T.R., Cook, A.Z, and Parker, P.E., 2010. Wind Speed Effects on the Quantity of *Xanthomas axonopodis* pv. *citr*i
3. Brandle, J.R. and Finch, S., How Windbreaks Work, University of Nebraska-Lincoln, Soil Conservation Service, EC-91-1763B.
4. Campbell, C.L, and Madden, L.V., 1990. Introduction to Plant Disease Epidemiology, John Wiley and Sons, New York.
5. Danos, E. Bonzazzola, R, Berger, R.D, Stall, R.E., and Miller, J.W., 1981. Progress of Citrus Canker on Some Species and Combinations in Argentina, Proc. Fla. State Hort. Soc., 94:1981,
6. Danos, E., Berger, R.D, and Stall, R. E., 1984, Temporal and Spatial Spread of Citrus Canker Within Groves, Phytopathology, Vol 74.

7. Fanci, L.J., 2001, The Disease Triangle: a plant pathological paradigm revisited, Online APS website, Teaching Articles.
8. Gottwald, T.R., McGuire, R.G., and Garran, S., 1988. Asiatic Citrus Canker: Spatial and Temporal Spread in Simulated New Planting Situations in Argentina, Vol. 76, No. 6.
9. Gottwald, T.R. and Graham, J.H., 1990. Spatial Pattern Analysis of Citrus Bacterial Spot in Florida Citrus Nurseries, 80:181-190.
10. Gottwald, T.R., Graham, J.H. and Engel, D.S., 1992. Analysis of foci of Asiatic Citrus Canker in a Florida Citrus Orchard, Plant Disease, 76: 386- 396.
11. Gottwald, T.R. and Timmer, L.W., July 1995. The efficacy of windbreaks in reducing the spread of citrus canker caused by *Xanthomonas campestris* pv. *citri*, Trop. Agric. (Trinidad).
12. Gottwald, T.R., Graham, J.H. and Schubert, T.S. , 1997, Citrus Canker in Urban Miami: An analysis of spread and prognosis for the future, Citrus Ind., 78, 72078.
13. Gottwald, T.R., Hughes, G., Graham, J.H., Sun, X., Riley, T., 2001. The Citrus Canker Epidemic in Florida: The Scientific Basis of Regulatory Policy for an Invasive Species, Phytopathology, Vol 91(1).
14. Gottwald, T.R., Sun, X., Riley, T. Graham, J.H., Ferrandino, F. and Taylor, E., 2002. Geo-Referenced Spatiotemporal Analysis of the Urban Citrus Canker Epidemic in Florida, Phytopathology, Vol 92, No. 4.
15. Gottwald, T.R. and Irey, M., 2007. Post Hurricane Analysis of Citrus Canker II: Predictive Model Estimation of Disease Spread and Area Potentially Impacted by Various Eradication Protocols Following Catastrophic Weather Events, Online, Plant Health Network, April 2007.
16. Gottwald, T.R., Bassanezi, R.B., Amorim, L., and Bergamin-Filho, A., 2007, Spatial Pattern Analysis of Citrus Canker- Infected Plantings in São Paulo, Brazil, and Augmentation of Infection Elicited by the Asian Leafminer, Phythopathology, 97:674-683.
17. Irey, M., Gottwald, T.R., Graham, J.H., Riley, T.D. and Carlton, G., 2006. Post-Hurricane Analysis of Citrus Canker Spread and Progress towards the Development of a Predictive Model to Estimate Disease Spread Due to Catastrophic Weather Events. Online, Plant Health Progress, August 2006.
18. Jeger, M.J, (Editor), 1989, Spatial Components of Plant Disease Epidemics, Prentice Hall, ISBN 0-13-824491-X.
19. Madden, L.V., Hughes, G., and van den Bosch, F., 2007. The Study of Plant Disease Epidemics, American Phytopathological Society, Minnesota. ISDN 978-0-89054-354-2.
20. Neri, F.M., Cook, A.R., Gibson, G.J., Gottwald, T.R. and Giligan, C.A., 2014. Bayesian Analysis for Inference of an Emerging Epidemic: Citrus Canker in Urban Landscapes, PLOS (Public Library of Sceince) Computational Biology, Vol 10, No 4.
21. Parker, P.E., Bock, C.H., Gottwald, T.R, 2005, Comparison of Techniques to Sample *Xanthomonas axonopodis* pv. *citri* in Windblown Spray, Plant Disease, Vol. 89, No 12.
22. Parnell, S., Gottwald, T.R., van der Bosch, F. and Gilligan, C.A., 2009. Optimal Strategies in the Eradication of Asian Citrus Canker in Heterogeneous Host Landscapes, Phytopathology, 99, 1370- 1376.

23. Pruvost, G., Boher, B., Brocherieux, C., and Chiroleu, F., 2002. Survival of *Xanthomonas axonopodis* pv. *citri* in Leaf Lesions and Simulated Splash Dispersal of Inoculum, Phytopathology, April 2002.
24. Ripley, B.D, 1988, Statistical Inference for Spatial Processes, Cambridge University Press.
25. Stall, R.E, Miller, J.W., Marco, G.M., de Echenique, B.I.C., 1980, Population Dynamics of Xanthomonas Citri causing Cancrosis of Citrus in Argentine, Proc. Fla. State Hort. Soc. 83: 1980.
26. Stall, R.E., Marco, G.M. and de Echenique, B.I.C, 1982, Importance of Mesophyll in Mature-Leaf Resistance to Cancrosis in Citrus, Phytopathology, Aug 1982.
27. Triola, M.F., 2005, Elementary Statistics, Ninth Edition, Pearson Education, Inc, Boston. Experimental and observational studies defined on page 20.
28. University of Minnesota, Sustainability Landscape Information Series (online series)
29. Vanderplank, J.E., 1963, Plant Diseases: Epidemics and Control, Academic Press, New York, 340 pp.
30. Verniere, C.J., Gottwald, T.R., Pruvost, O., 2003, Disease Development and Symptom Expression of *Xanthomonas axonopodis* pv. *citri* in Various Citrus Plant Tissues, Phytopathology, July 2003.
31. Venette, J.B., and Kennedy, B.W., 1975, Naturally produced aerosol of Pseudomonas cinae, Phytopathology, 65: 737-738.

Chapter 7: Field Study Investigation Summary

1. FDACS Press Release, 1998. Crawford Announces Changes to Canker Program, February 26, 1998 (Copy available on website).
2. Dixon, W.N., 2001. Citrus Canker Amendments Rule Development Workshop, Rule Chapter 5B-58. Oct 10, 2001, West Palm Beach, FL.
3. Florida Department of Agriculture and Consumer Services, Department of Plant Industries, Citrus Canker in Dade and Broward Cos. 1999, Report No. CCRAG-9, May 11, 1999, Minutes of Meeting. Submitted into evidence in Broward 17th District Court, in November 2000 (Case 00-18394 (08).
4. Garmin User Manual, GPS Meter for 12XL meter (available from the website).
5. Gaskalla, R., 2000, FDACS/DPI, Letter to David Lord, containing boundaries of study sites. December 22, 2000.
6. Gottwald, T.R, 2000, Presentation to the Broward Court Case 00-18394 (08) CACE. November 2000.
7. Gottwald, T.R., Hughes, G., Graham, J.H, Sun, X., Riley, T., 2001, The Scientific Basis of Regulatory Eradication Policy for an Invasive Species, Phytopathology, 91:30-34.
8. Gottwald, T.R., X. Sun, Riley, T. Graham, J.H., Ferrandino, F. and Taylor, E., 2002, Geo-Referenced Spatiotemporal Analysis of the Urban Citrus Canker Epidemic in Florida, Phytopathology, Vol 92, No. 4.
9. Madden, L.V., Hughes, G., and van den Bosch, F., 2007, The Study of Plant Disease Epidemics, American Phytopathological Society, Minnesota. ISDN 978-0-89054-354-2.

10. Neri F.M., Cook A.R., Gibson G.J., Gottwald T.R., Gilligan C.A., (2014) Bayesian Analysis for Inference of an Emerging Epidemic: Citrus Canker in Urban Landscapes. PLoS Comput Biol 0(4): e1003587. doi:10.1371/journal.pcbi.1003587
11. Lakeland Ledger, 1999. Canker Task Force is Girding for Backlash, Paul Power Jr., May 12, 1999.

Supplemental Notes

12. Citrus Canker Pathology Training Presentation, Citrus Health Response Program, Texas A+M, (provided on website).
13. Canteros, B.I., 2004, Citrus Canker in Argentina— A Review, Proc. Int. Soc. Citriculture, Paper No. 9, 2004.

Chapter 8: Undisclosed Studies

1. Department of Agriculture and Consumer Services v. Varela, 732 So. 2d 1146 (Fla. 3d DCA 1999)
2. FDACS/DPI, 1999, Citrus Canker in Dade and Broward Cos. Report No. CCRAG-9, May 11, 1999, Minutes of Meeting. Submitted into evidence in Broward 17th District Court, in November 2000 (Case 00-18394 (08).
3. FDACS website, "1900-ft Justification" statements. Various summaries of epidemiology studies were posted to the FDACS website. Also, a summary of the research leading to the 1900-ft rule was posted under "Frequently asked question," copies of these postings are provided on the supporting documents website.
4. FDACS, Commissioner Crawford's letter to President Clinton, July 24, 2000. Copy available on supporting documents website.
5. Gottwald,T.R and Graham, J. H., Research in Support of Eradication and Control of Asiatic Citrus Canker, Progress progress report 981-29. Copy available on supporting documents website.
6. Gottwald, T.R., Graham, J.H. and Schubert, T.S., 1997, An Epidemiology Analysis of the Spread of Citrus Canker in Urban Miami, Florida and Synergistic Interaction with the Asian Citrus Leaf Miner, Fruits, Vol 52-5, 371.
7. Gottwald, T.R., 1999, Canker Spread Study in Urban Miami. This document was submitted by FDACS in November 2000, to the Broward Court. The court termed the document an "interim report." The co-authors of the report are the same as in reference 7, but have been left off, because it is unknown if any of the co-authors had participated in the preparation of the document.
8. Gottwald, T.R, 2000, Presentation to the Broward Court Case 00-18394 (08) CACE. (copies of viewgraphs provided on supporting documents website).
9. Gottwald, T.R., Hughes, G., Graham, J.H, Sun, X., Riley, T., 2001, The Scientific Basis of Regulatory Eradication Policy for an Invasive Species, Phytopathology, 91:30-34.
10. Gottwald, T.R., Graham, J.H., Schubert, T.S., 2002. Citrus Canker, The Pathogen and its Impact, Plant Health Progress, online article available at www.apsnet.org.

11. Gottwald, T.R., X. Sun, Riley, T. Graham, J.H., Ferrandino, F. and Taylor, E., 2002, Geo-Referenced Spatiotemporal Analysis of the Urban Citrus Canker Epidemic in Florida, Phytopathology, Vol 92, No. 4.
12. Graham, J., Gottwald, T., Cubero, J., Drouillard, Survival of Xanthomonas campestris pv. citri on Various Surfaces, 2000, International Citrus Canker Research Workshop, June 20-22, 2000. Transcript only.
13. Graham, J.H., Gottwald, T.R., Cubero, J. and Achor, D. S., 2004, Xanthomonas axonopodis pv. citri: factors affecting successful eradication of citrus canker, Molecular Plant Pathology, 5(1), 1-15.
14. Wikipedia, Voronoi Diagrams, https://en.wikipedia.org/wiki/Voronoi_diagram1. Department of Agriculture and Consumer Services v. Varela, 732 So. 2d 1146 (Fla. 3d DCA 1999)

Chapter 9: A New History Emerges

1. CDC, Morbidity and Mortality Weekly Report, 1998. Surveillance for Acute Pesticide-Related Illness During Medfly Program, November 12, 1998.
2. Corneal v. State Plant Board, 95 So.2d 1 (Fla. 1957)
3. Dixon, W.N., 2001. Citrus Canker Amendments Rule Development Workshop, Rule Chapter 5B-58. Oct 10, 2001, West Palm Beach, FL.
4. FDACS - DPI, Comprehensive Report on Citrus Canker in Florida, December 2012.
5. Gottwald, T.R., Graham, J.H. and Schubert, T.S. , 1997, Citrus Canker in Urban Miami: An analysis of spread and prognosis for the future, Citrus Ind., 78, 72078.
6. Gottwald, T.R., Graham, J.H. and Schubert, T.S., 1997, An Epidemiology Analysis of the Spread of Citrus Canker in Urban Miami, Florida and Synergistic Interaction with the Asian Citrus Leaf Miner, Fruits, Vol 52-5, 371.
7. Gottwald, T.R, Sun, X., Riley, T., Graham, J. and Hughes, G., 1999. Citrus Canker Spread in Urban Miami, Internal Interim Report, sent to FDACS on October 13, 1999.
8. Gottwald, T.R., Hughes, G., Graham, J.H, Sun, X., Riley, T., 2001, The Scientific Basis of Regulatory Eradication Policy for an Invasive Species, Phytopathology, 91:30-34.
9. Gottwald, T.R., X. Sun, Riley, T. Graham, J.H., Ferrandino, F. and Taylor, E., 2002, Geo-Referenced Spatiotemporal Analysis of the Urban Citrus Canker Epidemic in Florida, Phytopathology, Vol 92, No. 4.
10. Gottwald, T.R., Graham, J.H., Schubert, T.S., 2002, Citrus Canker: The Pathogen and Its Impact, Plant Health Progress, published online at www.apsnet.org (official website of the American Phytopathological Society) .
11. Graham, J.H., 1998, "Citrus Canker, Control Efforts in Brazil, Prognosis for Florida", 1998, Citrus Industry, **79**(8), 54-57.
12. Graham, J.H., Gottwald, T.R., Cubero, J. and Achor, D. S., 2004. Xanthomonas axonopodis pv. citri: factors affecting successful eradication of citrus canker, Molecular Plant Pathology, 5(1), 1-15.

13. Madden, L.V., Hughes, G., and van den Bosch, F., 2007. The Study of Plant Disease Epidemics, American Phytopathological Society, Minnesota. ISDN 978-0-89054-354-2.
14. Naples Daily News, 1999. Gore Announces $25M in federal funds to fight citrus canker, March 13, 1999. Jennifer Maddox, Scripts Howard News Service.
15. Nielsen, K., 2000. Anatomy of a Quarantine, Miami New Times, July 6, 2000.
16. Office of Attorney General, 1997. Canker Fight Gets Major Funding Boost, April 18, 1997.
17. Orlando-Sun. 1997. Fear is Fallout of Spraying for Medflies, June 8, 1997.
18. Palm Beach Post, 2005. Canker program has citrus nurseries near 'collapse', Susan Salisbury, December 4, 2005.
19. Schubert, T.S. , Gottwald, T.R., Rizvi, S.A., Graham, J.H., Sun, X., Dixon, W.N., 2001, Meeting the Challenge of Eradicating Citrus Canker in Florida- Again, Plant Disease, Vol. 85-4.
20. USDA/APHIS, 1999, Citrus Canker Eradication Program: Environmental Assessment, April 1999. Agency contact listed as Dr. Stephen Poe, Program Officer.
21. Vanderplank, J.E., 1963, Plant Diseases: Epidemics and Control, Academic Press, New York, 340 pp.
22. Whiteside, J.O., "How Serious a threat is canker to Florida's Citrus Production, Citrus and Vegetable Journal, 1985.
23. Whiteside. J.O., September 1986, Citrus Canker, Some Facts, Speculations and Myths about this Highly Dramatized Bacterial Disease, Citrus and Vegetable Magazine.
24. Whiteside, J.O., April 1988, "The History and Rediscovery of Citrus Canker in Florida, Citrus and Vegetable Journal.

Supplemental Notes:

25. Gottwald, T.R., 2000. International Citrus Canker Research Workshop, June 21, 2000 (Day 2), Citrus Canker Epidemiology, Methodologies and Approaches. Transcript, Page 319.
26. Koizumi, M., Kimijima, E., Togawa, M., Masul, S., 1996. Dispersion of Citrus Canker Bacteria in Droplets and Prevention with Windbreaks, Proc. Int. Soc. Citriculture, 1996, Vol. 1.

Chapter 10: Post CCEP: Living with Canker

1. Bouffard, K., Canker Eradication Era Ends, Lakeland Ledger, October 6, 2006. (Article states that 65% of nursery stock had been destroyed, more than 80 thousand acres of groves destroyed, and 636 million dollars spent on compensation, and another 100 million would be budgeted to settle remaining claims).
2. Center for Biological Diversity, 2016. EPA Concludes Neonicotinoids Pose Risk to Bees, Fails to Analyze Other Pollinators, online publication.
3. Dewdney, M.M., Graham, J.H, 2014 Florida Citrus Pest Management Guide: Citrus Canker, PP-182, UF/IFAS, available on-line. .
4. Duan, Y., 2013, Prescription for Curing Citrus Greening: Apply Heat and Wait, Agricultural Research Magazine, USDA-ARS online publication, Aug 2013.

5. Evans, E.A., Ballen, F.H., Crane, J.H., 2014, Economic Potential of Producing Tahiti Limes in Southern Florida in the Presence of Citrus Canker and Citrus Greening, HortTechnology, Feb 2014, Vol. 24, No. 1, 95-106.
6. FDACS-DPI, Comprehensive Report on Citrus Canker in Florida, Revised December 2012 (Complete statistics on eradicated citrus tree by county)
7. Gottwald, T.R., 2007. Citrus Canker and Citrus Huanglongbing, Two Exotic Bacterial Diseases Threatening the Citrus Industry of the Western Hemisphere, Outlooks on Pest Management, October 2007. Online article, see link in online supporting documents website.
8. Hamilton,A., 2010, Squeezed: What You Don't Know About Orange Juice, Yale Agrarian Studies Series, Yale University Press.
9. Hunsberger, A., Gabel, K. and Mannion, C., 2006, University of Florida/ IFAS Extension, Citrus Leafminer (*Phyllocnistis citrella*). Short factsheet on CLM, which suggests neonicotinoid insecticides such as imidacloprid, to be used by growers.
10. International Union for Conservation of Nature (IUNC), 2014, Systemic Pesticides Pose Global Threat to Biodiversity and Ecosystem Services, June 24, 2014.
11. Jones, M.A, Stansly, P.A., Russo and Russo, J.M., Degree-day models will help growers time citrus leafminer sprays, April 15, 2013. (Research aimed at optimal timing of chemical sprays and use of synthetic pheromone based traps to control CLM based on knowledge of their life cycle).
12. Lowe, D., 2009, Current Situation, Management, and Economic Impact of Citrus Canker in Florida, USDA/APHIS/PPQ, Sept 2009 (online copy available).
13. Timmer, L.W. Biological Control of Citrus Leafminer Proves to be an IPM Success,reprinted on the web by permission of Florida Growers and Ranchers, Nov 1996.
14. USDA/APHIS, FDACS/DPI, 2006-2007, Citrus Health Response Plan (CHRP), developed jointly with representatives from USDA/ARS, UF/IFAS, Florida Citrus Mutual and the California Citrus Research Board.
15. USDA (2006). Evaluation of asymptomatic citrus fruit (Citrus spp.) as a pathway for the introduction of citrus canker disease (*Xanthomonas axonopodis* pv. citri)- March 2006. Raleigh, NC, U.S. Department of Agriculture, Animal and Plant Health Inspection Service, Plant Epidemiology and Risk Analysis Laboratory.
16. USDA/APHIS, 2007, Risk Management Analysis of Commercially Packed Citrus Fruit from Citrus Canker Disease Quarantine Area, Risk Management Analysis, June 2007.
17. USDA/APHIS, 2007, Risk Management Analysis of Commercially Packed Citrus Fruit from Citrus Canker Disease Quarantine Area, Risk Management Analysis, June 2007.
18. USDA/APHIS, 2008, An Updated Evaluation of Citrus Fruit (Citrus spp.) as a Pathway for the Introduction for the Introduction of Citrus Canker, Plant Epidemiology and Risk Analysis Laboratory.
19. USDA-National Agricultural Statistics Service (NASS), 2013 - 2014 Florida Citrus Summary.
20. US Environmental Protection Agency, Office of Chemical Safety and Pollution Prevention, Preliminary Pollinator Assessment to Support the Registration Review of Imidacloprid.

21. Whiteside, J.O., How Serious a threat is canker to Florida's Citrus Production, Citrus and Vegetable Journal, 1985.
22. Whiteside, J.O., 1988. The History and Rediscovery of Citrus Canker in Florida, Citrus and Vegetable Journal. April 1988.
23. White House Presidential Task Force, 2015. National Strategy to Promote the Health of Honey Bees and other Pollinators, Pollinator Health Task Force, May 19, 2015.
24. Wikipedia, Neonicotinoid. (summary of neonicotinoid usage and research on environmental impacts, which appears to be regularly updated. Contains 82 references, many to scientific studies).
25. Wikipedia, Imidacloprid (summary of basic facts of imidacloprid, usage and controversy on environmental impacts)

Florida Citrus Canker Technical Advisory Task Force Meeting Minutes

This list only includes meetings for which the minutes were requested and provided by FDACS.

General Meeting: May 14, 1999 - Presentation by Dr. Gottwald on Epidemiology Study
Science and Regulatory Issues Joint Meeting, June 30, 1999 - Vote by members on recommendation of 1900-ft rule.
General Meeting July 16, 1999 - Vote by members on recommendation
General Meeting, November 16, 1999- Presentation by Dr. Gottwald on citrus canker forecast model based on wind velocity and precipitation.

Other meetings: General Meeting: March 19, 1999, June 22, 1999, October 19, 1999, February 3, 2000, April 11, 2000.

Selected FDACS Press Releases

Feb 2, 1998, "Crawford Announces Changes to Canker Program"
Feb 12, 1999, "Crawford Appoints Panel to Combat Canker"
Jun 17, 1999, "Citrus Canker Eradication Project to Resume Cutting Exposed Trees"
Feb 2, 2000, "Crawford Unveils Bold New Canker Plan"
Nov 11, 2000, "Crawford Announces Program to Provide Free Citrus Trees"

www.ingramcontent.com/pod-product-compliance
Lightning Source LLC
Chambersburg PA
CBHW080652190526
45169CB00006B/2088